BURLEIGH DODDS SCIENCE: INSTANT INSIGHTS

NUMBER 79

Improving biosecurity in livestock production

I0130942

burleigh dodds
SCIENCE PUBLISHING

Published by Burleigh Dodds Science Publishing Limited
82 High Street, Sawston, Cambridge CB22 3HJ, UK
www.bdspublishing.com

Burleigh Dodds Science Publishing, 1518 Walnut Street, Suite 900, Philadelphia, PA 19102-3406, USA

First published 2023 by Burleigh Dodds Science Publishing Limited
© Burleigh Dodds Science Publishing, 2023. All rights reserved.

British Library Cataloguing in Publication Data
A catalogue record for this book is available from the British Library

ISBN 978-1-80146-631-8 (Print)
ISBN 978-1-80146-632-5 (ePub)

DOI: 10.19103/9781801466325

Typeset by Deanta Global Publishing Services, Dublin, Ireland

Contents

Series list

Title	Series number
Sweetpotato	01
Fusarium in cereals	02
Vertical farming in horticulture	03
Nutraceuticals in fruit and vegetables	04
Climate change, insect pests and invasive species	05
Metabolic disorders in dairy cattle	06
Mastitis in dairy cattle	07
Heat stress in dairy cattle	08
African swine fever	09
Pesticide residues in agriculture	10
Fruit losses and waste	11
Improving crop nutrient use efficiency	12
Antibiotics in poultry production	13
Bone health in poultry	14
Feather-pecking in poultry	15
Environmental impact of livestock production	16
Pre- and probiotics in pig nutrition	17
Improving piglet welfare	18
Crop biofortification	19
Crop rotations	20
Cover crops	21
Plant growth-promoting rhizobacteria	22
Arbuscular mycorrhizal fungi	23
Nematode pests in agriculture	24
Drought-resistant crops	25
Advances in detecting and forecasting crop pests and diseases	26
Mycotoxin detection and control	27
Mite pests in agriculture	28
Supporting cereal production in sub-Saharan Africa	29
Lameness in dairy cattle	30
Infertility/reproductive disorders in dairy cattle	31
Alternatives to antibiotics in pig production	32
Integrated crop–livestock systems	33
Genetic modification of crops	34

Chapter 1

On-farm strategies for preventing pig diseases: improving biosecurity

Jeroen Dewulf and Dominiek Maes, Ghent University, Belgium

1 Introduction

Pigs are susceptible to a wide range of diseases, including zoonotic infections, which can affect health, welfare and productivity, and thereby have a major economic impact. The implementation of biosecurity measures along the production chain is one of the major solutions to minimize the risk of introduction of these diseases into a farm, as well as their spread within the farm (Dewulf and Van Immerseel, 2018).

Recently, several studies have demonstrated a positive association between biosecurity and production parameters (Laanen et al., 2013; Postma et al., 2016) and between biosecurity and farm profitability (Corrégé et al., 2012; Siekkinen et al., 2012; Rojo-Gimeno et al., 2016; Collineau et al., 2017). Biosecurity has also been shown to have a positive impact on reducing the amount of antimicrobials used in pig production (Laanen et al., 2013; Postma et al., 2016). This is important considering that antimicrobial use in pig production has been identified as one of the highest among livestock sectors in European Union (EU) countries (Filippitzi et al., 2014; Carmo et al., 2017; Sarrazin et al., 2019).

Despite these documented associations and the recognized importance of biosecurity measures, there are still major shortcomings in the implementation of these measures in pig farming (Laanen et al., 2013; Backhans et al., 2015; Filippitzi et al., 2017). There are several examples of spread of infections due

http://dx.doi.org/10.19103/AS.2022.0103.10

to insufficient implementation of biosecurity measures, such as the spread of porcine epidemic diarrhea (PED) (Scott et al., 2016), the spread of highly pathogenic strains of porcine reproductive and respiratory syndrome virus (HP-PRRSV) (Brookes et al., 2015), the foot-and-mouth disease (FMD) epidemic in the United Kingdom in 2001 (Ellis-Iversen et al., 2011) and, most recently, the African swine fever (ASF) epidemic in Europe and Asia (Blome et al., 2020).

2 What is biosecurity?

Biosecurity consists of the combination of all measures implemented to reduce the risk of introduction and spread of disease agents (Dewulf and Van Immerseel, 2018). The measures to be established should not be seen as constraints but rather as part of a process aimed at improving the health of animals, people and the environment. Biosecurity can be subdivided into two main components:

1 External biosecurity, which is focused on keeping pathogens out of the herd; and
2 Internal biosecurity or bio-management, which is focused on preventing the spread of pathogens within the herd.

Implementing biosecurity requires the adoption of a set of attitudes and behaviors to reduce the risk in all activities involving animal production or animal care.

Infectious diseases may spread through many different transmission routes. Some are spread by airborne transmission whereas others are not; some pathogens may be transmitted through vectors while some may be spread through semen. Biosecurity measures generally aim at preventing these different transmission routes in an attempt to break the infection cycle. When designing biosecurity measures, one can approach the topic from the point of view of one specific pathogen and design measures that are specifically adapted to the epidemiology of that pathogen. Alternatively, a biosecurity plan can be made more generic and include the majority of the transmission routes with a focus on those that are more important, either because they are important in the transmission routes of many different pathogens or because they are of importance in the transmission pathways of the most prevalent or damaging diseases.

When designing biosecurity programs, there are some general principles that are of value in all settings:

1 Separation of high- and low-risk animals and environments;
2 Reduction of general infection pressure;

3 Recognition that not every transmission route is equally important;

4 Assessing risk as a combination of probability of transmission and frequency of occurrence of transmission routes; and

5 Recognizing that larger animal groups pose higher risks.

These principles are discussed in the following sections.

2.1 Separation of high- and low-risk animals and environments

To avoid pathogen transmission, it is important to try to keep the sources of infection (e.g. animals, persons, vermin), i.e. small animals or insects that may transmit pathogens, separated from susceptible contacts as much as possible. This can be achieved through the prevention of direct contacts between high-risk (infectious) and low-risk (susceptible) animals as well as the prevention of indirect contacts. Whenever contacts between high- and low-risk animals or compartments cannot be avoided (e.g. buying new breeding stock, visitors to the farm), precautionary measures should be implemented (e.g. quarantine period, changing of clothing and footwear, etc. – discussed later in the chapter).

2.2 Reduction of general infection pressure

Even with the best possible biosecurity, it is not possible, and probably not even desirable, to keep animals under sterile conditions. It is therefore very hard to prevent contact with potentially harmful infectious agents altogether. Biosecurity measures aim at reducing the infection pressure to below a level that allows the natural immunity of the animals to cope with infections. As a result, biosecurity is not about 'all or nothing', but it is about controlling the infection pressure as much as possible. For some specific pathogens, however (e.g. notifiable diseases or other pathogens such as *Brachyspira hyodysenteriae*), the aim is to fully prevent contact with the pathogen and its introduction into the farm.

2.3 Recognition that not every transmission route is equally important

Some transmission routes may transmit many different pathogens with a high efficiency (e.g. direct animal contact), whereas other transmission routes (e.g. feed) are of less importance (Dewulf et al., 2018) (Fig. 1). Therefore, when designing biosecurity control programs, it is important to focus first on high-risk transmission routes and only subsequently on lower-risk transmission routes. A frequently observed mistake in biosecurity is that emphasis is put on the wrong measures and much energy is put into the prevention of low-risk

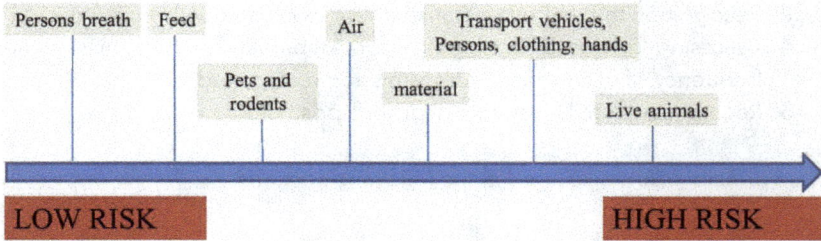

Figure 1 Theoretical ranking of different routes for infection transmission from low to high risk (after Boklund et al., 2008).

transmission routes whereas high-risk routes are ignored. This, of course, can depend on the pathogen involved, because some are specifically associated with so-called low transmission routes (e.g. some *Salmonella* serotypes with certain feed components).

2.4 Assessing risk as a combination of probability of transmission and frequency of occurrence of transmission routes

Besides the probability of transmission of a pathogen through different transmission routes, the frequency of occurrence of the transmission route is also important. If a certain transmission route (e.g. transmission via hands of animal care takers) has only a low probability of occurrence but handling is repeated very frequently (e.g. the animal care taker is touching the animals several times a day without precautionary measures), the risk of infection transmission ends up being very substantial. Therefore, when designing biosecurity programs, one also has to focus on those acts that are repeated very regularly as they, even if they carry only low risks, will still become important due to the frequency of occurrence.

2.5 Recognizing that larger animal groups pose higher risks

The larger the herds become, the more important biosecurity measures will be. This is firstly because large herds have more frequent contacts. There are more feed-delivery trucks coming to the herd, there are more frequent movements of animals, there are more professional visitors to the farm, etc. All of these carry a certain risk of infection transmission. Secondly, in large groups of animals, there are more animals that can become infected, maintain an infection cycle and build up infection pressure above the limit that the animals can cope with. It is often observed that farms gradually increase in size without fundamentally adapting their biosecurity measures. In these cases, the managers continue working according to their standards of the past and do not understand why they should adapt. The risks related to introduction and spread of an infection are, however, much more important in large herds.

3 External biosecurity measures

Based upon transmission routes, biosecurity measures have been developed, which aim to prevent either the introduction or spread of these pathogens in a pig herd. These measures can be subdivided into:

1 external biosecurity measures, aiming at prevention of pathogen introduction; and
2 internal biosecurity measures, aiming at prevention of within-herd spread of infections.

These are discussed in this and the following section.

In general terms, it can be stated that external biosecurity measures are mainly linked either to infrastructural aspects, such as organization of the buildings on the farm, measures to restrict entrance for animals and persons (e.g. hygiene lock, quarantine pen) or to measures affecting those outside the farm (e.g. entrance restrictions for visitors, hygiene of transport vehicles, safety of feed). External biosecurity measures include:

- purchasing policy;
- transport of animals, removal of manure and carcasses;
- supply of feed, water and equipment;
- access of personnel and visitors;
- vermin and bird control; and
- Location and environment.

In many cases, external biosecurity is better understood and implemented by farmers as compared to internal biosecurity. This is probably because external biosecurity measures have received more attention in the past in control of epidemic diseases.

3.1 Purchasing policy

The introduction of non-proprietary animals or genetic material (e.g. semen) might lead to the introduction of pathogens for which there is no farm immunity. Pathogen transmission occurs very effectively via direct contact between infected and susceptible animals (Filippitzi et al., 2017). Therefore, the importance of biosecurity in purchasing policy is high in protecting a farm from many pathogens. From a biosecurity viewpoint, the primary aim should be to avoid the purchase of animals or genetic material as much as possible (Amass and Baysinger, 2006; Dewulf, 2014; Filippitzi et al., 2017). A fully closed herd or production system, that is, a system where no other animals are purchased, has a substantially lower risk of pathogen introduction. Moreover, besides the risk of pathogen introduction, the frequent addition of 'naïve' animals may also favor the continuous circulation of herd-specific pathogens. This may hamper the control and eradication of certain pathogens in a herd.

However, avoiding the introduction of new animals in modern pig production is often very difficult or unwelcome because farms want to keep up with the genetic progress made by breeding companies. In addition, some farmers do not have the facilities, management skills or time to raise their own breeding gilts. The male offspring of breeding sows (production of new breeding gilts) on the farm generally also have a lower value compared to the offspring from the other sows (production of fattening pigs). Bernaerdt et al. (2021) showed that 57% of the farms were purchasing breeding gilts in Belgium. Garza-Moreno et al. (2017) described, based on 321 questionnaires completed by 118 pig veterinarians from 18 countries, that approximately half of the surveyed farms purchased replacement gilts. Therefore, whenever new animals are introduced, a number of precautionary measures should be taken.

3.1.1 Limit the frequency of introduction

Both the frequency of introduction and the number of purchased animals will influence the risk of pathogen introduction (Fèvre et al., 2006; Laanen et al., 2013). In both cases, the key principle from a biosecurity viewpoint is 'the less, the better'. However, sometimes it can be advisable to increase the size of the group of purchased animals (e.g. new gilts) as this may reduce the frequency of new introductions. It is generally less risky to buy 20 gilts five times a year rather than 10 gilts ten times a year.

3.1.2 Limit the number of sources

It is very important to limit the number of source herds for animals and AI centers for semen as much as possible (Dewulf, 2014). Several studies have shown that introducing animals from different source herds increases the risk of disease introduction (Hege et al., 2002; Lo Fo Wong et al., 2004). Source herds should have a documented high health status (Pritchard et al., 2005; Kirwan, 2008; Dewulf, 2014). This status may include the certified absence of a number of infectious diseases (e.g. specific pathogen-free status), which avoids the unintended introduction of new pathogens in the receptor herd (Laanen et al., 2010; Filippitzi et al., 2017).

3.1.3 Respect good quarantine

Newly purchased animals should always be introduced first in a quarantine stable. A good quarantine stable is fully isolated from the other animal facilities and should be accessed through a separate entry with its own hygiene lock (Fig. 2). During the quarantine period, animals should be clinically inspected to confirm that no signs of any diseases are present. Animals can also be sampled to detect any current infections and to assess the immune status of animals. The quarantine period also allows implementation of acclimatization procedures, for example, vaccinating the newly introduced animals to assure a sufficient level of immunity when they are brought into contact with the resident animals (Barceló and Marco, 1998; Corrégé, 2002; Pritchard et al., 2005; Calvar et al., 2012; Dewulf, 2014). A quarantine period should last at least 4 weeks; however, for some diseases longer periods are recommended (PRRSV and PCV2, 6-8 weeks; *Mycoplasma hyopneumoniae*, 8-10 weeks) (Eijck, 2003; Pritchard et al., 2005). Besides quarantine, vaccination and/or acclimatization may be required even earlier in the life of gilts (Pieters and Fano, 2016). In a study by Bernaerdt et al. (2021), 95% of the farms that purchased breeding gilts practiced a quarantine period. Comparable results were obtained in Sweden (Backhans et al., 2015). On most of these farms, the quarantine was located on the farm site. The median duration of the quarantine period was 42 days. The most common acclimatization practice was vaccination, although in some farms exposure of gilts to farm-specific microorganisms was also undertaken.

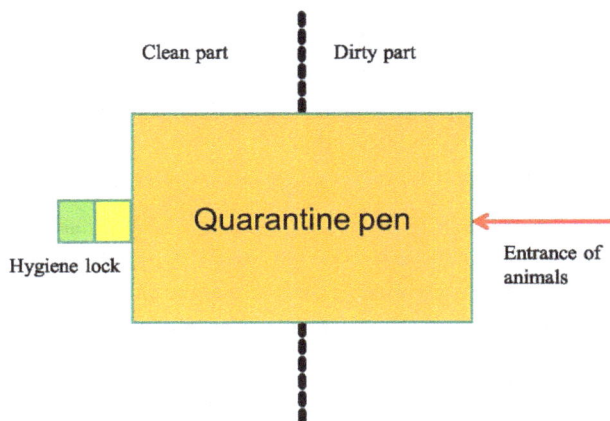

Figure 2 Design of a good quarantine stable.

3.2 Transport of animals, removal of manure and carcasses

Pathogens can spread through the transport of live animals and/or the removal of carcasses or manure. This spreading of infectious germs can be achieved directly (via secreta and excreta of diseased animals and carcasses) or indirectly (via fomites or carcasses, the rendering truck, people and their material, rodents, domestic animals and manure).

3.2.1 Use clean animal-transport vehicles

Epidemiological field studies have identified contagious livestock lorries as a main source of contamination for many disease-causing agents, including Classical swine fever (CSF) virus (Fritzemeier et al., 2000), *M. hyopneumoniae* (Hege et al., 2002), *Actinobacillus pleuropneumoniae* (Fussing et al., 1998 and Hege et al., 2002), *Brachyspira hyodysenteriae* (Windsor and Simmons, 1981) and *Salmonella* (Rajkowski et al., 1998). Lorries collecting livestock should always be empty, cleaned and disinfected and dry before entering the premises (Pritchard et al., 2005; Dewulf, 2014). Although this is a well-known principle, this is not sufficiently respected in practice. Very cold weather conditions may, for example, hamper thorough cleaning and disinfection. It is also the case that trucks that collect culled breeding sows are not always empty upon arrival on the farm. It is sometimes required that a lorry for the transport of livestock should be empty for at least a couple of hours or days before it can enter the farm. This might provide additional risk reduction. However, it is clear that thorough cleaning, disinfection and drying are the principal biosecurity measures. These cannot be replaced in favor of 'downtime', when lorries are empty. Pigs that have been in contact with the lorry during loading may not be moved back to the stable in order to minimize the chance of introducing pathogens through an insufficiently cleaned lorry. For the same reason, the lorry-driver should not be allowed to enter the stables. A loading bay is advisable, and, if available, it should also be cleaned and disinfected after each movement of animals (Pritchard et al., 2005; Backhans et al., 2015).

3.2.2 Ensure separation between the clean and the dirty area

The principle of the clean and dirty 'road' in a pig farm means that there is a clear separation between the clean and the dirty (risky) sections of the premises (Hémonic et al., 2010; Anonymous, 2010; Neumann, 2012; Filippitzi et al., 2017). All inbound and outbound traffic involving external parties (e.g. feed supply, movement of liquid manure, external transport of animals, etc.) are always directed via the 'dirty road'. The 'clean road' is restricted to within-farm movements of animals and goods and potentially the supply of safe products, but only in fully cleaned and disinfected lorries.

Only the 'dirty road' is relatively easily accessible to visitors, suppliers and consumers. The collection of carcasses is for obvious reasons part of the dirty section. Barrels, wheelbarrows and other tools used for this purpose may only be returned to the clean section after they have been thoroughly cleaned and disinfected.

Liquid manure should always be conveyed via the dirty road. Furthermore, it is advisable to the farmer to use farm-specific discharge pipes in order to prevent pipes from a manure removal company, which recently have been in contact with manure on other farms, from also being used on the farm. Some farms also have a manure processing plant on or nearby the pig stables. It is obvious that all this movements relating to manure should take place via the dirty road.

Recent studies have indicated that the clean-dirty area principle is not completely respected by manure removal and supply companies in a number of EU countries (Filippitzi et al., 2017). This means that farmers should ensure that clean-dirty areas are clearly defined with signs indicating how to adhere to these areas. The use of 'nudging' might also help promote the right behavior in these circumstances (e.g. Reddy et al., 2017).

3.2.3 Management of carcasses

Carcasses should always be considered as a major source of infectious material. Animals often die due to an infection and will thereby potentially spread a lot of infectious material. Carcasses should therefore be removed from pens and the stables as soon as possible. They should be stored in a well-insulated place (Meroz and Samberg, 1995; Pritchard et al., 2005) that cannot be accessed by vermin (rodents, insects, etc.) as they could spread the infectious material. After the collection of carcasses, the cadaver storage room should be thoroughly cleaned and disinfected. The persons handling carcasses should always wear disposable gloves for their own safety as well as to avoid further spread of pathogens (Pritchard et al., 2005; Filippitzi et al., 2017).

The cadaver storage room should be located in a place where the rendering company can collect the carcasses without entering the farm to avoid pathogen introduction through these potentially risky channels (Evans and Sayer, 2000; McQuiston et al., 2005; Pritchard et al., 2005; Anonymous, 2010). It is also recommended to have a cooled cadaver storage room for hygienic reasons (to reduce the growth of pathogenic or spoilage bacteria as well as reduce odors). Such cooled rooms generally have a higher storage capacity, making it possible to reduce the frequency of visits by the rendering company (Fig. 3).

3.3 Supply of feed, water and equipment

Feed itself should generally not pose a risk due to the strict hygienic conditions of its production. However, swill feeding (banned for decades under EU law)

Figure 3 Cooled cadaver storage close to the public road.

is a practice that has previously been associated with large outbreaks of infectious diseases, including CSF (Horst et al., 1997; Fritzemeier et al., 2000). In the recent ASF outbreak in Asia, swill feeding was also considered to be one of the major drivers of virus spread (Blome et al., 2020).

The quality of drinking water for pigs often leaves much to be desired. The water may originate from different sources (rainwater, wells, etc.) after which it is typically stored in a reservoir and then distributed to the animals. The water can be contaminated in the reservoir or in pipes leading to the drinking nipples used by pigs, and biofilms may be formed. Regular examination of the drinking water quality (at least once a year) both at the source (prior to entering the stables) and at the drinking nipples in the stables is therefore advisable. Proper treatment procedures for water and/or the pipes should be implemented in case of microbiological contamination.

The introduction of all sorts of equipment, such as floating panels or shovels, which come into contact with animals and their manure may also introduce pathogens. Therefore, it is advisable to avoid the introduction of new equipment as much as possible or, if introduced, it should be disinfected first.

3.4 Access of personnel and visitors

Humans can act as a vector for pathogens if they have been in contact with infected animals and subsequently come in to contact with susceptible animals without taking any preventive measures. This type of transmission has been proven through experiments with several pathogens, including the transmissible gastroenteritis (TGE) virus (Alvarez et al., 2001), *Escherichia coli* (Amass et al., 2003) and CSFV (Ribbens et al., 2007). The transmission occurs mainly through particles of excreta from infected animals on footwear and clothing. There is also the possibility of biological transmission of pathogens between persons and pigs that can infect both, such as the H1N1 influenza virus (Wentworth et al., 1997) or *methicillin-resistant Staphylococcus aureus* (MRSA) ST398 (Huijsdens et al., 2006). The first measure to be taken is therefore to limit the number of persons that can access the stables to the absolute minimum. This requires measures to prevent visitors entering accidentally (Fig. 4).

When visitors and personnel enter the stables, they should always wear clean, herd-specific clothes and footwear. They should also wash their hands properly (Pritchard et al., 2005; Hémonic et al., 2010; Dewulf, 2014) (Fig. 5). This latter is a simple and very useful measure, but one which is often neglected. Pathogens are efficiently transferred via the hands of animal care takers through direct contact with the animals (Vangroenweghe et al., 2009; Hémonic et al., 2010; Backhans et al., 2015). A study by Lo Fo Wong et al. (2004) has shown that the chance of testing positive for *Salmonella spp.* is reduced as a result of consistent hand washing before entering a section with pigs.

To ensure a change of clothing and washing of hands by personnel and visitors, a good hygiene lock should be in place (Vangroenweghe et al., 2009)

Figure 4 Clear entrance control at the farm to avoid unwanted visitors.

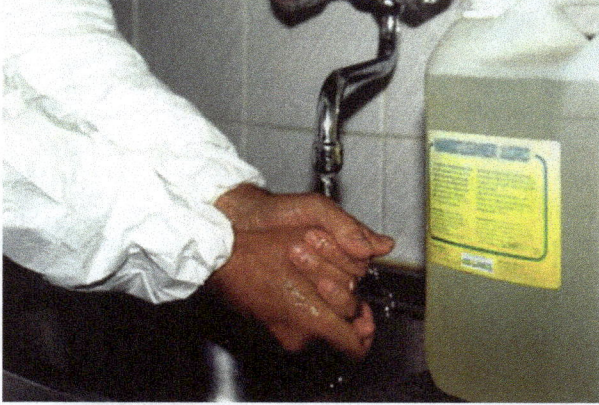

Figure 5 Proper washing of hands.

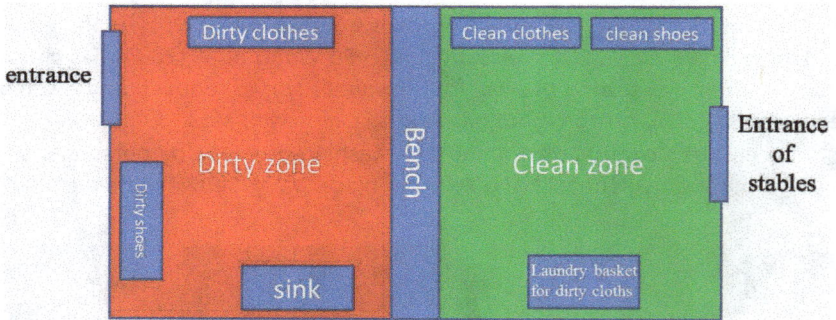

Figure 6 Design of a hygiene lock.

(Fig. 6). In this hygiene lock, a clear physical separation (e.g. bench) between the dirty and the clean area should be provided.

When entering the hygiene lock, the following steps should be followed:

1 Take off your jacket and shoes.
2 Wash your hands with disinfecting soap.
3 Step over the bench and put on a clean overall and boots.
4 Disinfect the boots with the boot washer before entering the stable.
5 When returning to the hygiene lock, clean and disinfect the boots with the boot washer.
6 Put the boots on the appropriate shelf or in a disinfectant bath.
7 Take off the dirty overall and put it in the laundry basket.
8 Step over the bench and wash your hands before you put on your own jacket and shoes again.

9 In farms with high health standards, visitors and personnel are often obliged to shower before entering the farm. The main benefit of this requirement is to ensure that all potentially contaminated clothing will be replaced by farm-specific clothing and that the hands are washed thoroughly. In addition, it discourages unnecessary visits and increases awareness of the need to implement proper biosecurity measures (Moore, 1992; Amass and Clark, 1999).

Often, a quarantine period of 24 or even 48 h, during which no contact with pigs can occur, is also required for visitors before they can have access to the farm. This is based on the argument that germs excreted by pigs can survive on humans for a specific period. During this period, persons could passively excrete germs and transfer them to susceptible animals. There is little evidence in the scientific literature that this is a significant risk. The authors are aware of only one study from 1970, which found that FMDV could be isolated from the nose and mouth of people who had been in contact with animals infected with the FMDV. On one person, the virus could still be isolated 28 h after contact with the infected pigs. After 48 h, this was no longer possible (Sellers et al., 1970). If all required precautionary measures are taken as described above, this downtime period probably has little additional value. It is clear that, whatever downtime is used, it is no substitute for the hygiene measures on entering a farm described here.

3.5 Vermin and bird control

A number of pathogens can be transmitted from outside the farm or between different areas within the farm directly or indirectly by rodents, birds, dogs and cats. They may act as reservoirs for herd-specific pathogens that will continue to circulate in the farm (Andres and Davies, 2015). Rodents and birds can also cause damage to equipment and farm buildings or contaminate feed if they can access it (Backhans and Fellstrom, 2012).

To control vermin, an efficient control program is required. This is often developed in collaboration with specialized companies (Lister, 2008; Hémonic et al., 2010; Dewulf, 2014; Backhans et al., 2015; Filippitzi et al., 2017). It is important to prevent vermin living in close proximity to stables. This can be achieved by removing potential hiding places near to the stables (e.g. plants, piles of dirt, etc.). Feed should be stored in closed containers with no access for rodents or birds (Lister, 2008; Anonymous, 2010). The entrance of birds into stables can be prevented by covering all air inlets with nets (Loncke and Dewulf, 2018). Pets should also be kept out of the stables as they can be vectors for pathogens. The use of cats or dogs to control rats and mice is therefore not advisable (Vangroenweghe et al., 2009).

3.6 Location and environment

The location of the farm and the density of pig farming in the immediate area are important factors for airborne and vector-borne disease transmission. The presence of wildlife in the neighborhood of pig farms may also pose certain risks.

3.6.1 Airborne transmitted diseases

Bacterial transmission through the air is particularly important over short distances (<2 km), hence the importance of distance to the nearest neighbor. Rose and Madec (2002) concluded that the number of farms within a range of 2 km significantly increased the frequency of respiratory disorders in a farm. In the case of airborne transmission of *M. hyopneumoniae*, the distance to neighboring farms proved to be the most important factor (Goodwin, 1985; Dee et al., 2009). Maes et al. (2000a) reported increased prevalence of *M. hyopneumoniae*, swine influenza viruses and Aujeszky's disease virus in herds located in areas with a high density of pig farming. Mintiens et al. (2003) showed that the combination of the distance to a neighboring farm and pig herd density in an area are major risk factors for the spread of CSFV. When the building of a new pig farm is being planned, the distance to the nearest neighboring pig farm can be a determining factor in choosing a location. The predominant direction of the wind could be taken into account as well. Knowledge of the presence of diseases in neighboring farms is equally important. The spread of liquid manure originating from other farms should also be avoided in the neighborhood of the farm.

It has become possible to equip stables with high-performance air filtration systems that can block airborne pathogens. This might be worth considering in densely-populated livestock areas, especially if animals in the farm are free of endemic diseases in the area. Filtration of incoming air, in combination with standard biosecurity procedures, has been demonstrated to prevent transmission of PRRSV in susceptible herds (Alonso et al., 2013). The authors showed that air filtration reduced the risk of introduction of novel PRRSV strains by approximately 80%, indicating that on large sow farms with good biosecurity in pig-dense regions, approximately four out of five PRRSV outbreaks may be attributable to aerosol transmission. Air filtration may also be helpful to prevent airborne transmission of other pathogens, such as swine influenza virus and *M. hyopneumoniae*.

3.6.2 Wild animals

Direct or indirect contact with wild boars may cause transmission of pathogens (e.g. CSFV (Fritzemeier et al., 2000), Aujeszky's disease virus (Artois et al., 2002) and ASFV (Blome et al., 2020)). In regions where free ranging wild boars are present, it is therefore important to keep wild animals out of the farm with a solid fence (Amass and Clark, 1999) with an underground depth of 30 to 40 cm (Hartung, 2005). Even if pigs are kept indoors, wild boars should not be able to come near the farm in order to avoid the indirect transmission of pathogens (e.g. airborne transmission, through vectors, through contact with stored feed, etc.).

4 Internal biosecurity measures

Internal biosecurity measures include:

- Compartmentalization, working lines and equipment;
- Management of diseases;
- Measures during the farrowing and suckling period;
- Management of the nursery and fattening unit; and
- Cleaning and disinfection.

These are discussed in the following sections.

4.1 Compartmentalization, working lines and equipment

Animals of different ages may have different levels of susceptibility to specific pathogens. It is therefore crucial to keep different age groups separate and to work in a well-defined sequence. Equipment and materials (e.g. bedding material, feeders, drinking troughs, boots, spades, syringes, needles, etc.) may also play an important role in the transmission of a large number of diseases.

4.1.1 Working lines and separate hygiene locks

An important basic rule to prevent the spread of infections between different age groups is establishing and maintaining working lines within the farm. A fixed routine is created that is always used to visit and work in the stables. During rounds to the stables, the youngest animals are visited first, followed by the pregnant sows, the older age groups, the quarantine and sick animals and, finally, the cadaver storage unit. For each age category, and especially for risk-bearing groups (e.g. quarantine stables, sickbay), it is recommended to provide an additional hygiene lock allowing proper changing of clothing, footwear and washing of hands.

4.1.2 Equipment in the various compartments

Designated equipment should be provided for each working line. A brush, a shovel or floating panels can easily be contaminated with feces that contain a large number of germs. It is therefore recommended to use dedicated equipment in each section. Equipment has to be clearly recognizable (e.g. using different colors) (Fig. 7) to avoid moving it from one section to another (Vangroenweghe et al., 2009; Laanen, 2011; Gelaude et al., 2014). The same rule applies to clothing and footwear for the same reason.

4.1.3 Boot washers and disinfection baths

To avoid germs on footwear, boot washers and disinfection baths can be placed between the different production units. Effective disinfection can only be achieved if dirt and feces are removed from boots in advance. This can be done with a boot washer and water (preferably adding a detergent). The boots then have to be placed in a visually clean solution with a disinfectant. This protocol requires that the concentration of the disinfectant and the duration of contact must follow manufacturer instructions (Amass et al., 2000). Disinfection baths that are not used properly will inadvertently increase the number of germs on the boots. This wastes both time and money and can even increase the risk of disease spread. However, it is not practical to stand for many minutes in a disinfection bath before being able to go to another section. This problem can be avoided by providing an extra pair of boots at each disinfection bath, so there's always a pair of boots ready to be used at each bath while the other boots are placed in the disinfecting solution. The presence of foot baths also draws the attention of staff and visitors to the importance of biosecurity on the farm (Amass et al., 2000).

Figure 7 Floating panels in different colors to be used in different compartments and to avoid using the same material in different age groups.

4.2 Management of diseases

Disease management concerns all actions related to the correct handling and treatment of diseased animals. This includes proper diagnostics, isolation of sick animals and disease registration as well as improvement of the immunity status of susceptible animals, in particular through vaccination. Good herd heath management should result in a good understanding of the specific health status of the herd and in use of appropriate preventive treatments to avoid disease and its subsequent losses.

4.2.1 Sickbay

Diseased pigs should be isolated in a sick bay in order to protect other animals from exposure to pathogens through infected excretions and secretions. A good sickbay is fully separated from the other animals, preferably in a separate building (Hémonic et al., 2010; Dewulf, 2014). Once an animal has been in the sick bay, it should not return to the regular stables as it is very likely that it will transmit any remaining pathogens to healthy animals. The sick bay should also be accessed separately by farm workers and the necessary hygienic measures (e.g. changing of coverall, footwear, washing hands, etc.) should be implemented when entering and leaving. The sick bay should ideally only be visited at the end of the working round (Vangroenweghe et al., 2009; Backhans et al., 2015).

4.2.2 Use of needles and medicines

There is extensive literature on the spread of germs via injection equipment (needles and syringes) (Hémonic et al., 2010; Filippitzi et al., 2017). In practice, the same needles are often used on individual pig farms and only replaced when they become blunt! These needles may get contaminated with bacteria through use and storage (Fig. 8). In addition, when injecting sick animals, the needle (and consequently the bottle) can become contaminated with a pathogen. Injecting multiple animals with the same needle thus increases the risk of spreading germs.

Ideally, single use needles should be used (Hémonic et al., 2010). If this is not feasible, it is highly advisable to use one needle for each group (e.g. litter or pen). Using the same needles for different age groups should be avoided and needles should be replaced before they become blunt, both for hygiene and animal welfare reasons. Previously opened bottles should be stored in a hygienic environment at the recommended environmental conditions such as temperature and light. Needle-free injection devices, often used for vaccination, may also help to decrease the risk of iatrogenic infection transmission.

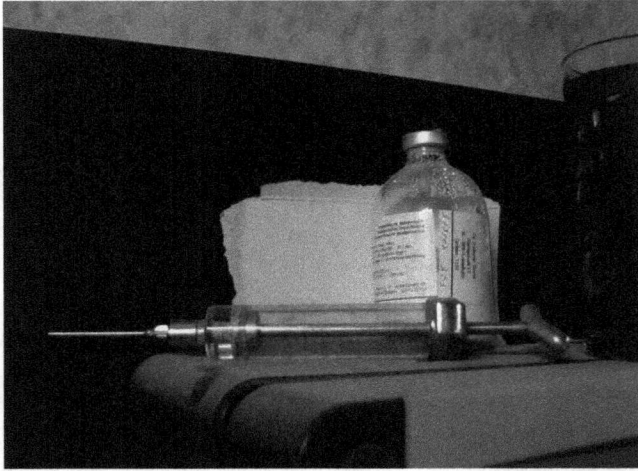

Figure 8 Improper stored needles, syringes and medicines may be a source of infection.

4.2.3 Returning to younger age group

In case of slower growth rate of some piglets compared to the remainder of the group, it is important to avoid moving more slowly growing piglets to a batch of younger piglets. These slower growing piglets may be suffering from one or more infectious diseases. By transferring these pigs to a younger age group, possible carriers of germs are brought into a susceptible population (Vangroenweghe et al., 2009; Dewulf, 2014; Filippitzi et al., 2017). When it is expected that a piglet has a low probability of reaching slaughter age, euthanasia should be considered. This might be a better option both for the piglet, for welfare reasons, and for the whole litter, because their health is protected. If euthanasia is not believed to be the right option, then these animals should be isolated in the sickbay until they can be transported to the slaughterhouse.

4.3 Measures during the farrowing and suckling period

Pathogens can be transmitted from sows to offspring via the placenta or via contaminated colostrum and milk. They can also be transmitted by direct contact, for example, through the nose, skin, nipples and the udder. Cross-fostering in particular increases the risk of pathogen transmission from infected or carrier sows to susceptible piglets. This is particularly the case if cross-fostered piglets do not have (sufficient) maternal antibodies, for example, of PRRSV, as indicated by Zimmerman et al. (2012). Another route of pathogen transmission in the farrowing unit is the repeated use of equipment or materials (e.g. castration blades, elastrator for tail docking, ear-tagger, iron injection needles) from one piglet to another without intermediate cleaning and disinfection.

4.3.1 Washing the sows

Before sows are placed in the farrowing pen, they should be dewormed and washed in order to prevent the transmission of germs from the sow gestation unit to the farrowing unit. Washing of sows should be done before they enter the farrowing pen. If it is done in the farrowing unit, then contaminated aerosols are generated and spread over the entire farrowing unit. This significantly decreases the effect of washing and is contrary to the basic principle of using separate hygiene measures for different age groups or stables in a farm.

4.3.2 Cross-fostering

Mixing litters in the farrowing pen is a significant way for spread of infection to different animal groups. Sows that carry *Streptococcus suis* can infect their piglets during or immediately after parturition (Amass et al., 1996). Further spread of *S. suis* may take place if piglets are moved to other litters. This principle applies to other germs as well. If 5% of piglets are moved between pens more than 48 hours after birth, there is an increased probability of problems with PRRS (Duinhof et al., 2006). From a biosecurity viewpoint, it is advisable to avoid cross-fostering as much as possible. If this is not possible, it is recommended to limit cross-fostering to one occasion and to perform it shortly after birth.

4.3.3 Equipment for treatment of the piglets

Equipment that is used in the farrowing pen, such as blades for castration, is exposed to secretions and excretions from piglets. Germs can then be transferred from one piglet to another. It is necessary to clean and disinfect equipment when using it for a different piglet (immerse in disinfectant). This limits the chance of disease transmission (Vangroenweghe et al., 2009; Filippitzi et al., 2017).

4.4 Management of the nursery and fattening unit

4.4.1 All-in/all-out

The all-in/all-out (AI/AO) principle helps to prevent cross-contamination between consecutive production batches and makes it possible to clean and disinfect stables thoroughly between different production batches. The strict application of this AI/AO principle is a very important measure to break the infection cycle from one production batch to the next (Clark et al., 1991). It is fundamentally important that the stable is empty at the end of a production batch. Sometimes only 95% of a batch is removed with a few (light weight) animals kept in the stables and mixed with animals in the next batch. Although few, these animals are very likely a source of infection for the next groups (as discussed earlier). Moving piglets back to younger age groups is also a risky practice since it can also bring pathogens to a susceptible population. Weaners are a vulnerable age group due to their temporary lower immune status, the higher presence of diverse pathogens in this life period (Johnson et al., 2012) and the risk of fighting when pigs are mixed (Cameron, 2012). When moving animals from one production stage to the next (e.g. from the farrowing house to the nursery pen), it is advisable to keep the groups together as much as possible rather than sorting all animals according to weight. The latter will result in a lot of mixing which substantially increases the likelihood of spread of infections (Maes et al., 2008; Hémonic et al., 2010).

4.4.2 Stocking density

A high stocking density induces stress, which results in an increased susceptibility to infections and an increased excretion of germs. Many infected pigs in a small area can cause a sharp rise in infection pressure. Various studies have shown that a higher stocking density in different production phases increases the occurrence of respiratory as well as digestive tract disorders (Pointon et al., 1985; Maes et al., 2000a,b; Stärk, 2000; Laanen, 2011). In addition, it has been shown repeatedly that there is a positive connection between available space per animal and its daily growth (Dewulf et al., 2007). In many cases recommended stocking densities, as prescribed in EU legislation, are based on outdated research and have not evolved with recent advances in the industry. It is therefore recommended to consider these norms as the absolute minimum requirements and not as target values (Dewulf et al., 2007). Studies have shown that the optimal values for stocking density are on average 20 to 24 % above EU requirements (Hamilton et al., 2003; Laanen, 2011).

4.5 Cleaning and disinfection

Pens, feeding troughs and equipment contaminated by feces can maintain an infection cycle because new animals continue to become infected. These animals will consequently secrete germs and will re-infect the environment. To break the infection cycle between consecutive litters, a thorough cleaning and disinfection (C&D) of pens is required.

A complete C&D protocol consists of seven steps (Vangroenweghe et al., 2009a; Hémonic et al., 2010; Laanen, 2011; Dewulf, 2014; Luyckx, 2016; Van Immerseel et al., 2018):

1 Dry cleaning to remove all organic material;
2 Soaking of all surfaces preferably with detergent;
3 High pressure cleaning with water to remove all dirt. This step will be much easier, faster and more effective if a good soaking step is performed beforehand;
4 Drying of the stable to avoid dilution of the disinfectant applied in the next step;
5 Disinfection of the stable to achieve further reduction of the concentration of germs;
6 Drying of the stable to assure that animals afterwards cannot come into contact with pools of remaining disinfectant; and
7 Testing the efficiency of the procedure through sampling of the surface.

With the aid of pressure plates, it is possible to simply and rapidly check all surfaces for microbial contamination. These plates measure and quantify the presence of bacterial contamination. The results are expressed in colony-forming units (CFU) per plate. When all these steps are performed correctly, an additional empty period to further reduce the infection load is not required (Luycks, 2016).

5 Measuring biosecurity

'You need to be able to measure, to be able to improve' is one of the most famous quotes of William Thomson (better known as Lord Kelvin), a famous British scientist of the nineteenth century. This is certainly true for biosecurity and hygiene. The inability to measure accurately and reproducibly the biosecurity and hygiene status of a farm has long been one of the main obstacles to improving safety. If farm managers are to enhance the biosecurity or hygiene status of their farm, it is essential to provide them with quantitative goals and benchmarks relating to the biosecurity and hygiene status of a farm. This makes it possible to identify measures for improvement and assess their effectiveness, if possible quantitatively.

Several systems have been developed to provide an inventory of biosecurity measures for animal production. They are often developed as checklists or as manuals either by independent advisory organizations or as supporting material in the marketing of specific disease prevention products (e.g. vaccines). An example of the latter is the COMBAT system developed by the pharmaceutical company Boehringer Ingelheim that helps to identify

Table 1 Overview of different biosecurity scoring tools or checklists with some of their characteristics

System	Developer/owner	Disease focus	Language	Risk-based scoring	Benchmarking	Availability	Weblink
COMBAT	Boehringer Ingelheim	PRRS	Multiple languages	Yes	Yes	Free	https://www.prrs.com/disease-control/control/combat
Biocheck.Ugent	Ghent University	Generic	Multiple languages	Yes	Yes	Free	https://biocheck.ugent.be/en
CFSPH Biosecurity	Iowa State University	Generic	English	No	No	Free	https://www.cfsph.iastate.edu/biosecurity/
APIQ√	Australian Pork	Generic	English	No	No	Free	https://www.apiq.com.au/

biosecurity hazards in relation to the PRRSV infections in pig production. Like the COMBAT PRRS system, many of these systems were developed from the point of view of the control of a specific disease. In the United States, there is also the more generic biosecurity evaluation tool developed by the Center for Food Security and Public Health at Iowa State University. This system consists of a series of checklists for different components of biosecurity. Other countries have published fact sheets on the principles of biosecurity, for example to reduce the risk of introduction and spread of CSFV. In Australia, information on biosecurity is also available for pig producers via the Australian Pork Industry (2003). More recently, they have also developed manuals for pig producers, including an internal audit system called APIQ√® which stands for Australian Pork Industry Quality Assurance Program. This system enables producers to demonstrate that their on-farm practices reflect good farming practice for management, animal welfare, food safety, biosecurity and traceability.

At Ghent University, a risk-based biosecurity scoring system (Biocheck. UGent™) has been developed to quantify on-farm biosecurity (Laanen et al., 2010). It does not focus on a specific disease but rather approaches biosecurity in general and focuses on those aspects that are common for the transmission of many different types of infectious diseases. The Biocheck.UGent™ system consists of a number of questions divided into several subcategories for internal and external biosecurity. Depending on the importance of a particular biosecurity measure, the score per question is multiplied by a weight factor (Laanen et al., 2013; Gelaude et al., 2014). The subcategories also have a specific weight factor corresponding to their relative importance in disease transmission. The Biocheck.UGent™ scoring system thus provides a risk-based score that takes into account the relative importance of each biosecurity measure. The Biocheck.UGent™ scoring tool is accessible to everybody and is free of charge (www.biocheck.ugent.be). After filling in the questionnaire, the results allow evaluation of strong and weak points in the biosecurity on a farm and provide a basis for improvement. Besides the systems discussed here, there are likely to be others, with a focus on a specific country or region, which are unknown to the authors or inaccessible due to language limitations. Table 1 provides an overview of the systems described in this section.

6 Conclusion

As discussed in this chapter, biosecurity in pig production is a combination of many different measures aimed at preventing the introduction and spread of pathogens in a farm. It forms the basis of any disease control program. Both external and internal biosecurity measures are of utmost importance both to prevent introduction of pathogens on a farm and to avoid infection spread between animal populations in a farm. Although most of the measures to be implemented are logical and generally easy to apply, it requires strong discipline to adhere to the measures in daily farm practice. Yet, those who do surely will see the benefits.

7 Where to look for further information

More information can be found in the book: Biosecurity in Animal Production and Veterinary Medicine, from principles to practice. Edited by: Jeroen Dewulf, Ghent University, Belgium, Filip Van Immerseel, Ghent University, Belgium

Also the website www.biocheck.ugent.be contains tons of tips and tricks on biosecurity.

8 References

Alonso, C., Murtaugh, M. P., Dee, S. A. and Davies, P. R. (2013). Epidemiological study of air filtration systems for preventing PRRSV infection in large sow herds. *Preventive Veterinary Medicine* 112(1-2), 109-117.

Alvarez, R. M., Amass, S. F., Stevenson, G. W., Spicer, P. M., Anderson, C., Ragland, D., Grote, L., Dowell, C. and Clark, L. K. (2001). Evaluation of biosecurity protocols to prevent mechanical transmission of transmissible gastro-enteritis virus of swine by pork production unit personnel. *Pig Journal* 48, 22-33.

Amass, S. F. and Baysinger, A. (2006). Swine disease transmission and prevention. In: Straw, B. E., Zimmerman, J. J., D'Allaire, S. and Taylor, D. J. (Eds), *Diseases of Swine* (9th edn). Blackwell Publishing Ltd., Oxford, UK, 1075-1098.

Amass, S. F. and Clark, L. K. (1999). Biosecurity considerations for pork production units. *Journal of Swine Health and Production* 7, 217-228.

Amass, S. F., Clark, L. K., Knox, K., Wu, C. C. and Hill, M. A. (1996). *Streptococcus suis* colonization of piglets during parturition. *Journal of Swine Health and Production* 4, 269-272.

Amass, S. F., Halbur, P. G., Byrne, B. A., Schneider, J. L., Koons, C. W., Cornick, N. and Ragland, D. (2003). Mechanical transmission of enterotoxigenic *Escherichia coli* to weaned pigs by people, and biosecurity procedures that prevented such transmission. *Journal of Swine Health and Production* 11, 61-68.

Amass, S. F., Vyverberg, B. D., Ragland, D., Dowell, C. A., Anderson, C. D., Stover, J. H. and Beaudry, D. J. (2000). Evaluating the efficacy of boot baths in biosecurity protocols. *Journal of Swine Health and Production* 8, 169-173.

Andres, V. M. and Davies, R. H. (2015). Biosecurity measures to control *Salmonella* and other infectious agents in pig farms: a review. *Comprehensive Reviews in Food Science and Food Safety* 14(4), 317-335.

Anonymous (2010). Food and agriculture organization of the United Nations/world organisation for animal health/world bank. Good practices for biosecurity in the pig sector - issues and options in developing and transition countries. *FAO Animal Production and Health Paper No. 169*.

Artois, M., Depner, K. R., Guberti, V., Hars, J., Rossi, S. and Rutili, D. (2002). Classical swine fever (hog cholera) in wild boar in Europe. *Revue Scientifique et Technique* 21(2), 287-303.

Backhans, A. and Fellstrom, C. (2012). Rodents on pig and chicken farms – a potential treath to human and animal health. *Infection Ecology and Epidemiology* 2, 17093.

Backhans, A., Sjölund, M., Lindberg, A. and Emanuelson, U. (2015). Biosecurity level and health management practices in 60 Swedish farrow-to-finish herds. *Acta Veterinaria Scandinavica* 57, 14.

Barceló, J. and Marco, E. (1998). On farm biosecurity. Proceedings of the 15th IPVS Congress, Birmingham, England, 5–9 July, 129–133.

Bernaerdt, E., Dewulf, J., Verhulst, R., Bonckaert, C. and Maes, D. (2021). Purchasing policy, quarantine and acclimation practices of breeding gilts in Belgian pig farms. *Porcine Health Management* 7(1), 25.

Blome, S., Franzke, K. and Beer, M. (2020). African swine fever – a review of current knowledge. *Virus Research* 287, 198099.

Boklund, A., Barfod, K., Mortensen, S., Houe, H. and Uttenthal, Å. (2008). Exotic diseases in swine: evaluation of biosecurity and control strategies for classical swine fever.

Brookes, V. J., Hernández-Jover, M., Holyoake, P. and Ward, M. P. (2015). Industry opinion on the likely routes of introduction of highly pathogenic porcine reproductive and respiratory syndrome into Australia from South-East Asia. *Australian Veterinary Journal* 93(1–2), 13–19.

Calvar, C., Heugebaert, S., Caille, M. E. and Roy, H. (2012). La quarantaine. Des préconisations de techniciens diversifies. Des conduits multiples chez de très bons éleveurs. *Rapport d'étude, Chambres d'agriculture de Bretagne*.

Cameron, R. (2012). Integumentary system: skin, hoof, and claw. In: Zimmermann, J. J., Karriker, L. A., Ramirez, A., Schwartz, K. J. and Stevenson, G. W. (Eds), *Diseases of Swine*. Wiley-Blackwell, Chichester, West Sussex, UK, 926–991.

Carmo, L. P., Schuepbach-Regula, G., Muentener, C., Chevance, A., Moulin, G. and Magouras, I. (2017). Approaches for quantifying antimicrobial consumption per animal species based on national sales data: a Swiss example (2006–2013). *Eurosurveillance* 22(6), 30458.

Clark, L., Freeman, M., Scheidt, A. and Knox, K. (1991). Investigating the transmission of Mycoplasma hyopneumoniae in a swine herd with enzootic pneumonia. *Veterinary Medicine* 86, 543–550.

Collineau, L., Rojo-Gimeno, C., Léger, A., Backhans, A., Loesken, S., Nielsen, E. O., Postma, M., Emanuelson, U., Beilage, E. G., Sjölund, M., Wauters, E., Stärk, K. D. C., Dewulf, J., Belloc, C. and Krebs, S. (2017). Herd-specific interventions to reduce antimicrobial usage in pig production without jeopardising technical and economic performance. *Preventive Veterinary Medicine* 144, 167–178.

Corrégé, I. (2002). Problématique de l'introduction des reproducteurs. *Techniporc* 25, 27–30.

Corrégé, I., Fourchon, P., Le Brun, T. and Berthelot, N. (2012). Biosécurité et hygiène en élevage de porcs: état des lieux et impact sur les performances technico-économiques. *Journées Recherche Porcine* 44, 101–102.

Dee, S., Otake, S., Oliveira, S. and Deen, J. (2009). Evidence of long distance airborne transport of porcine reproductive and respiratoy syndrome virus and Mycoplasma hyopneumoniae. *Veterinary Research* 40, 1.

Dewulf, J. (2014). An online risk-based biosecurity scoring system for pig farms. *Veterinary Ireland Journal* 4, 426–429.

Dewulf, J., Tuyttens, F., Lauwers, L., Van Huylenbroeck, G. and Maes, D. (2007). De invloed van de hokbezettingsdichtheid bij vleesvarkens op productie, gezondheid en welzijn. *Vlaams Diergeneeskundig Tijdschrift* 76, 410–416.

Dewulf, J. and Van Immerseel, F. (2018). General principles of biosecurity in animal production and veterinary medicine. In: Dewulf, J. and Van Immerseel, F. (Eds), *Biosecurity in Animal Production and Veterinary Medicine, from Principle to Practice*. ACCO (vol. 2018), 63–76.

Dewulf, J., Postma, M., Vanbeselaere, B., Maes, D. and Filippitzi, M. E. (2018). Transmission of pig diseases and biosecurity in pig production. In: Dewulf, J. and Van Immerseel, F. (Eds), *Biosecurity in Animal Production and Veterinary Medicine, from Principle to Practice*. ACCO (vol. 2018), 295–328.

Duinhof, T., Van de Ven, S. C. G. and Van Schaik, G. (2006). A survey among veterinarians and pig farmers in the Netherlands: more focus needed on diagnostic approach and on-farm contact structures in the control of PRRS. Proceedings of the 19th IPVS Congress, Copenhagen, Denmark (vol. 1), 234.

Eijck, I. A. J. M. (2003). *Gezond starten, gezond blijven. Praktijkonderzoek veehouderij*. Animal Science Group, Wageningen.

Ellis-Iversen, J., Smith, R. P., Gibbens, J. C., Sharpe, C. E., Dominguez, M. and Cook, A. J. (2011). Risk factors for transmission of foot-and-mouth disease during an outbreak in southern England in 2007. *Veterinary Record* 168(5), 128.

Evans, S. J. and Sayer, A. R. (2000). A longitudinal study of Campylobacter infection of broiler flock in Great Britain. *Preventive Veterinary Medicine* 46(3), 209–223.

Fèvre, E. M., Bronsvoort, B. M. C., Hamilton, K. A. and Cleaveland, S. (2006). Animal movements and the spread of infectious diseases. *Trends in Microbiology* 14(3), 125–131.

Filippitzi, M. E., Brinch Kruse, A., Postma, M., Sarrazin, S., Maes, D., Alban, L., Nielsen, L. R. and Dewulf, J. (2017). Review of transmission routes of 24 infectious diseases preventable by biosecurity measures and comparison of the implementation of these measures in pig herds in six European countries. *Transboundary and Emerging Diseases* 65(2), 381–398. doi: 10.1111/tbed.12758.

Filippitzi, M. E., Callens, B., Pardon, B., Persoons, D. and Dewulf, J. (2014). Antimicrobial use in pigs, broilers and veal calves in Belgium. *Vlaams Diergeneeskundig Tijdschrift* 83(5), 215–224.

Fritzemeier, J., Teuffert, J., Greiser-Wilke, I., Staubach, Ch., Schlüter, H. and Moennig, V. (2000). Epidemiology of classical swine fever in Germany in the 1990s. *Veterinary Microbiology* 77(1-2), 29–41.

Fussing, V., Barfod, K., Nielsen, R., Møller, K., Nielsen, J. P., Wegener, H. C. and Bisgaard, M. (1998). Evaluation and application of ribotyping for epidemiological studies of Actinobacillus pleuropneumoniae in Denmark. *Veterinary Microbiology* 62(2), 145–162.

Garza-Moreno, L., Segalés, J., Pieters, M., Romagosa, A. and Sibila, M. (2017). Survey on *Mycoplasma hyopneumoniae* gilt acclimation practices in Europe. *Porcine Health Management* 3, 21.

Gelaude, P., Schlepers, M., Verlinden, M., Laanen, M. and Dewulf, J. (2014). Biocheck.UGent: a quantitative tool to measure biosecurity at broiler farms and the relationship with technical performances and antimicrobial use. *Poultry Science* 93(11), 2740–2751.

Goodwin, R. F. W. (1985). Apparent reinfection of enzootic-pneumonia-free pig herds: search for possible causes. *The Veterinary Record* 116(26), 690–694.

Hamilton, D. N., Ellis, M., Wolters, B. F., Schninckel, A. P. and Wilson, E. R. (2003). The growth performance of the progeny of two swine sire lines reared under different floor space allowances. *Journal of Animal Science* 81(5), 1126–1135.

Hartung, J. (2005). Zur Abschirmung von Beständen aus tierhygienischer Sicht. *Deutsche Tierärztliche Wochenschrift* 112, 313–316.

Hege, R., Zimmermann, W., Scheidegger, R. and Stärk, K. D. C. (2002). Incidence of reinfections with *Mycoplasma hyopneumoniae* and *Actinobaillus pleuropneumoniae* in pig farms located in respiratory-disease-free regions of Switzerland – identification and quantification of risk factors. *Acta Veterinaria Scandinavica* 43(3), 145–156.

Hémonic, A., Corrégé, I. and Lanneshoa, M. (2010). Quelles sont les pratiques de bioséccurité et d'hygiène en élevages de porcs? *Techni-Porc.* 33, 7–13.

Horst, H. S., Huirne, R. B. and Dijkhuizen, A. A. (1997). Risks and economic consequences of introducing classical swine fever into the Netherlands by feeding swill to swine. *Revue Scientifique et Technique* 16(1), 207–214.

Huijsdens, X. W., van Dijke, B. J., Spalburg, E., van Santen-Verheuvel, M. G., Heck, M. E., Pluister, G. N., Voss, A., Wannet, W. J. and de Neeling, A. J. (2006). Community-acquired MRSA and pig-farming. *Annals of Clinical Microbiology and Antimicrobials*. A.K. Press, Johnson 5, 26.

Johnson, A. K., Edwards, L. N., Niekamp, S. R., Philips, C. E., Sutherland, M. A., Torrey, S., Casey-Trott, T., Tucker, A. L., Widowski, T. (2012). Behavior and welfare. In: Zimmermann, J. J., Karriker, L. A., Ramirez, A., Schwartz, K. J. and Stevenson, G. W. (Eds), *Diseases of Swine*. Wiley-Blackwell, Chichester, West Sussex, UK, 180–249.

Kirwan, P. (2008). Biosecurity in the pig industry - an overview. *Cattle Practice* 16, 147–154.

Laanen, M. (2011). Waarop letten voor meer bioveiligheid? *Varkensbedrijf* 11, 30–33.

Laanen, M., Beek, J., Ribbens, S., Vangroenweghe, F., Maes, D. and Dewulf, J. (2010). Bioveiligheid op varkensbedrijven: ontwikkeling van een online scoresysteem en de resultaten van de eerste 99 deelnemende bedrijven. *Vlaams Diergeneeskundig Tijdschrift* 79, 302–306.

Laanen, M., Persoons, D., Ribbens, S., de Jong, E., Callens, B., Strubbe, M., Maes, D. and Dewulf, J. (2013). Relationship between biosecurity and production/antimicrobial treatment characteristics in pig herds. *Veterinary Journal* 198(2), 508–512.

Lister, S. A. (2008). Biosecurity in poultry management. In: Patisson, M., McMullin, P. F., Bradbury, J. M. and Alexander, D. J. (Eds), *Poultry Diseases* (6th edn.). Saunders Elsevier, Beijing, China, 48–65,

Lo Fo Wong, D. M. A., Dahl, J., Stege, H., van der Wolf, P. J., Leontides, L., von Altrock, A. and Thorberg, B. M. (2004). Herd-level risk factors for subclinical *Salmonella* infection in European finishing-pig herds. *Preventive Veterinary Medicine* 62(4), 253–266.

Loncke, T. and Dewulf, J. (2018). Rodent control in animal production. In: Dewulf, J. and Van Immerseel, F. (Eds), *Biosecurity in Animal Production and Veterinary Medicine, from Principle to Practice*. ACCO (vol. 2018), 283–294.

Luyckx, K. (2016). Evaluation and implication of cleaning and disinfection of broiler houses and pig nursery units. Phd thesis. Ghent University. Faculty of Veterinary Medicine, Merelbeke, Belgium. Available at: https://biblio.ugent.be/publication/8081692.

Maes, D., Deluyker, H., Verdonck, M., Castryck, F., Miry, C., Vrijens, B. and de Kruif, A. (2000a). Herd factors associated with the seroprevalences of four major respiratory pathogens in slaughter pigs from farrow-to-finish pig herds. *Veterinary Research* 31(3), 313–327.

Maes, D., Deluyker, H., Verdonck, M., Castryck, F., Miry, C., Vrijens, B., Ducatelle, R. and de Kruif, A. (2000b). Noninfectious herd factors associated with macroscopic and microscopic lung lesions in slaughter pigs from farrow to-finish pig herds. *The Veterinary Record* 148, 41–46.

Maes, D., Segales, J., Meyns, T., Sibila, M., Pieters, M. and Haesebrouck, F. (2008). Control of Mycoplasma hyopneumoniae infections in pigs. *Veterinary Microbiology* 126(4), 297–309.

McQuiston, J. H., Garber, L. P., Porter-Spalding, B. A., Hahn, J. W., Pierson, F. W., Wainwright, S. H., Senne, D. A., Brignole, T. J., Akey, B. L. and Holt, T. J. (2005). Evaluation of risk factors for the spread of low pathogenicity H7N2 avian influenza virus among commercial poultry farms. *Journal of the American Veterinary Medical Association* 226(5), 767–772.

Meroz, M. and Samberg, Y. (1995). Disinfecting poultry production premises. Revue scientifique et technique. *Revue Scientifique et Technique.* Office International des Epizooties 14(2), 273–291.

Mintiens, K., Laevens, H., Dewulf, J., Boelaert, F., Verloo, D. and Koenen, F. (2003). Risk analysis of the spread of classical swine fever virus through 'neighbourhood infections' for different regions in Belgium. *Preventive Veterinary Medicine* 60(1), 27–36.

Moore, C. (1992). Biosecurity and minimal disease herds. *Veterinary Clinics of North America. Food Animal Practice* 8(3), 461–474.

Neumann, E. J. (2012). Disease transmission and biosecurity. In: Zimmerman, J. J., Karriker, L. A., Ramirez, A., Schwartz, K. J. and Stevenson, G. W. (Eds), *Diseases of Swine* (10th edn.). John Wiley & Sons Inc., Iowa, 141–164.

Pieters, M. and Fano, E. (2016). *Mycoplasma hyopneumoniae* management in gilts. *Veterinary Record* 178(5), 122–123.

Pointon, A. M., Heap, P. and McCloud, P. (1985). Enzootic pneumonia of pigs in South Australia–factors relating to incidence of disease. *Australian Veterinary Journal* 62(3), 98–101.

Postma, M., Backhans, A., Collineau, L., Loesken, S., Sjölund, M., Belloc, C., Emanuelson, U., Grosse Beilage, E., Stärk, K. D. C. and Dewulf, J. (2016). The biosecurity status and its associations with production and management characteristics in farrow-to-finish pig herds. *Animal* 10(3), 478–489.

Pritchard, G., Dennis, I. and Waddilove, J. (2005). Biosecurity: reducing disease risks to pig breeding herds. *In Practice* 27(5), 230–237.

Rajkowski, K. T., Eblen, S. and Laubauch, C. (1998). Efficacy of washing and sanitizing trailers used for swine transport in reduction of *Salmonella* and *Escherichia coli*. *Journal of Food Protection* 61(1), 31–35.

Reddy, S. M. W., Montambault, J., Masuda, Y. J., Keenan, E., Butler, W., Fisher, J. R. B., Asah, S. T. and Gneezy, A. (2017). Advancing conservation by understanding and influencing human behavior. *Conservation Letters* 10(2), 248–256.

Ribbens, S., Dewulf, J., Koenen, F., Maes, D. and De Kruif, A. (2007). Evidence of indirect transmission of classical swine fever virus through contacts with people. *The Veterinary Record* 160(20), 687–690.

Rojo-Gimeno, C., Postma, M., Dewulf, J., Hogeveen, H., Lauwers, L. and Wauters, E. (2016). Farm-economic analysis of reducing antimicrobial use whilst adopting improved management strategies on farrow-to-finish pig farms. *Preventive Veterinary Medicine* 129, 74–87.

Rose, N. and Madec, F. (2002). Occurrence of respiratory disease outbreaks in fattening pigs: relation with the features of a densely and a sparsely populated pig area in France. *Veterinary Research* 33(2), 179–190.

Sarrazin, S., Joosten, P., Van Gompel, L., Luiken, R. E. C., Mevius, D. J., Wagenaar, J. A., Heederik, D. J. J., Dewulf, J. and EFFORT Consortium (2019). Quantitative and qualitative analysis of antimicrobial usage patterns in 180 selected farrow-to-finish pig farms from nine European countries based on single batch and purchase data. *Journal of Antimicrobial Chemotherapy* 74(3), 807–816.

Scott, A., McCluskey, B., Brown-Reid, M., Grear, D., Pitcher, P., Ramos, G., Spencer, D. and Singrey, A. (2016). Porcine epidemic diarrhea virus introduction into the United States: root cause investigation. *Preventive Veterinary Medicine* 123, 192–201.

Sellers, R. F., Donaldson, A. I. and Herniman, K. A. J. (1970). Inhalation, persistence and dispersal of foot-and-mouth disease virus by man. *Journal of Hygiene* 68(4), 565–573.

Siekkinen, K. M., Heikkilä, J., Tammiranta, N. and Rosengren, H. (2012). Measuring the costs of biosecurity on poultry farms: a case study in broiler production in Finland. *Acta Veterinaria Scandinavica* 54, 12.

Stärk, K. D. (2000). Epidemiological investigation of the influence of environmental risk factors on respiratory disease in swine. *Veterinary Journal* 159(1), 37–56.

Vangroenweghe, F., Ribbens, S., Vandersmissen, T., Beek, J., Dewulf, J., Maes, D. and Castryck, F. (2009). Keeping Pigs Healthy (in Dutch). In: Vangroenweghe, F., Print, D. C. L. and Signs (Eds), (1st edn). Zelzate, Belgium.

Van Immerseel, F., Luyckx, K., De Reu, K. and Dewulf, J. (2018). Cleaning and disinfection. In: Dewulf, J. and Van Immerseel, F. (Eds), *Biosecurity in Animal Production and Veterinary Medicine, from Principle to Practice*. ACCO (vol. 2018), 133–158.

Wentworth, D. E., McGregor, M. W., Macklin, M. D., Neumann, V. and Hinshaw, V. S. (1997). Transmission of swine influenza virus to humans after exposure to experimentally infected pigs. *Journal of Infectious Diseases* 175(1), 7–15.

Windsor, R. S. and Simmons, J. R. (1981). Investigation into the spread of swine dysentery in 25 herds in East Anglia and assessment of its economic significance in five herds. *The Veterinary Record* 109(22), 482–484.

Zimmerman, J. J., Benfield, D. A., Dee, S. A., Murtaugh, M. P., Stadejek, T., Stevenson, G. W. and Torremorell, M. (2012). Porcine reproductive and respiratory syndrome virus. In: Zimmermann, J. J., Karriker, L. A., Ramirez, A., Schwartz, K. J. and Stevenson, G. W. (Eds), *Diseases of Swine*. Wiley-Blackwell, Chichester, West Sussex, UK, 1675–1777.

Chapter 2

Disease identification and management on the pig farm

Dominiek Maes, Jeroen Dewulf, Filip Boyen and Freddy Haesebrouck, Ghent University, Belgium

1 Introduction

2 Disease identification

3 Disease management and control: overview

4 External biosecurity

5 Internal biosecurity

6 Vaccination and antimicrobial medication

7 Future trends in diagnostics and disease monitoring and control

8 Conclusion

9 Where to look for further information

10 References

1 Introduction

Despite the importance of animal health, many pig herds worldwide are permanently endemically infected with important pathogens. Examples of such pathogens include porcine reproductive and respiratory syndrome virus (PRRSV), porcine circovirus type 2 (PCV-2), *Mycoplasma hyopneumoniae*, *Actinobacillus pleuropneumoniae*, *Haemophilus parasuis*, pathogenic *Escherichia coli* strains, *Brachyspira* spp. and *Streptococcus* spp. They cause tremendous economic losses to the pig sector. Unfortunately, during the last decades, pig health has not paralleled the improvement levels as observed in pig production and reproductive performance. Prevalence figures of lung lesions in slaughter pigs for instance are comparable to those of 20 years ago (Meyns et al., 2011). This urges for proper control measures to improve the health of pigs.

Endemic diseases are highly prevalent, even in professionally managed pig herds. This is not surprising because many facultative pathogenic bacteria are part of the microbiota and because possible sources of infection and transmission routes are very numerous, for example, sow-to-piglet, pig-to-pig,

http://dx.doi.org/10.19103/AS.2017.0013.17

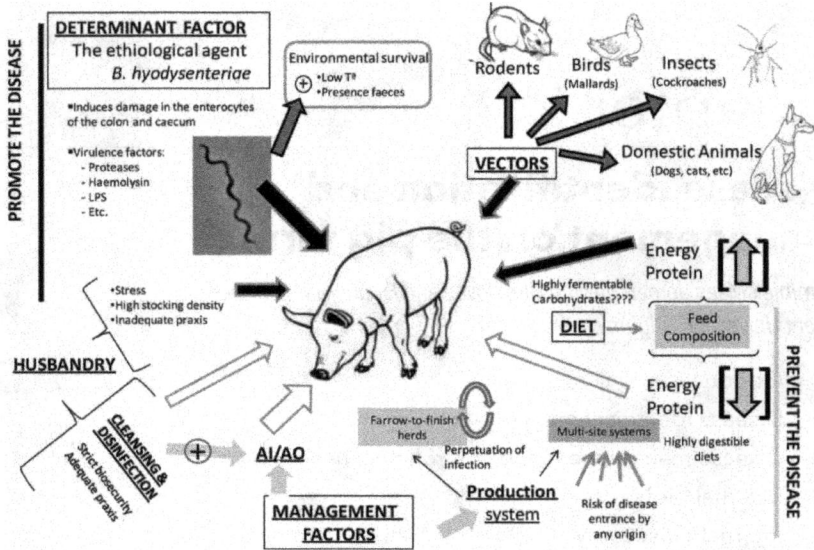

Figure 1 Factors influencing the occurrence and transmission of *Brachyspira hyodysenteriae* (Alvarez-Ordóñez et al., 2013).

via semen, aerosol, airborne transmission in pig-dense areas, people, rodents, insects, domestic and feral non-swine animals, birds, fomites, carcasses and vehicles (Amass and Baysinger, 2006). Also, often a large number of animals share the same airspace, or are raised on the same location, facilitating the transmission of pathogens. An example of the numerous transmission routes is shown in Fig. 1 for *Brachyspira hyodysenteriae* infections.

In intensive pig production systems, endemic diseases often result from an interaction of different infectious agents, and are influenced by environmental, nutritional and management conditions. Consequently, many pig diseases are multifactorial, with the clinical outcome, the severity of lesions and economic losses being the result of an interplay between pathogens, the animal host defence and the environment. The many interactions between these factors are often not fully known. Therefore, many pig diseases are called 'syndromes' or 'complexes', indicating that the precise aetiology and/or pathogenesis are not fully understood (Table 1).

Endemic pig diseases constitute a major threat for a profitable and sustainable pig production worldwide. They lead to decreased animal performance and carcass quality, more infections with zoonotic agents in humans, increased antimicrobial use, less animal welfare, reduced work satisfaction (and motivation) of the pig producer and finally a poorer image for the pig sector. Not only clinical disease but also subclinical infections, for example, with PRRSV, PCV2, swine influenza, *M. hyopneumoniae* and *A.*

Table 1 Examples of pig diseases called 'syndromes' or 'complexes', indicating that the exact aetiology and/or pathogenesis are not fully understood

Porcine reproductive and respiratory syndrome	Postpartum dysgalactia syndrome
Periparturient failing to thrive syndrome	Haemorrhagic bowel syndrome
Post-weaning multisystemic wasting syndrome	Porcine stress syndrome
Porcine dermatitis and nephropathy syndrome	Necrotic ear syndrome
Porcine ulcerative dermatitis syndrome	Porcine respiratory disease complex
Gastric ulceration	

pleuropneumoniae do have a measurable negative effect on performance. Decreases in daily weight gain in fattening pigs due to such subclinical infections typically range from 15 to 40 g (Rohrbach et al., 1993; Regula et al., 2000; Maes et al., 2003; Díaz et al., 2012).

With clinical disease, production losses are considerably higher (Allan and Ellis, 2000). However, in general, the number of clinically affected herds or animals remains low. Consequently, although decreased performance per pig is lower in subclinically than in clinically infected animals, at a population level, and from an economic perspective, the impact of subclinical infections to the pig sector is likely larger than the effect of clinical disease. Clinically diseased pigs are the tip of the iceberg, and while these animals obviously require appropriate treatment, addressing the potentially widespread subclinical infection 'that lies beneath' is essential (Maes et al., 2012).

Holtkamp et al. (2007) ranked and quantified productivity and economic losses in the United States due to the major swine health challenges based on a survey including 19 large production companies. The veterinarians were asked 1) to rank all significant health challenges in the company over the last year, 2) to provide an estimate of the percentage of animals that were affected annually and 3) to estimate the losses in affected herds for each ranked health challenge. The results for the finishing and nursery units are shown in Fig. 2 and 3, respectively. Estimated losses associated with the four most important health challenges are shown in Table 2.

Niemi et al. (2016) reviewed 130 publications to estimate the cost of production diseases in pigs. Reviewed studies showed great case-by-case variation in the costs of production diseases. Economic losses due to porcine respiratory disease complex were about €6.8 per fattening pig produced by an affected herd. Depending on pathogen and case, these losses ranged from €2 to €19 per fattening pig produced. In the studies analysed, the reduction in returns due to pre-weaning mortality was between €12 and €23 per litter, and due to post-weaning mortality between €2 and €4 per pig. The costs of postpartum dysgalactia syndrome amounted to €95 per affected sow. The costs of lameness in sows ranged from €145 to €180 per lame sow.

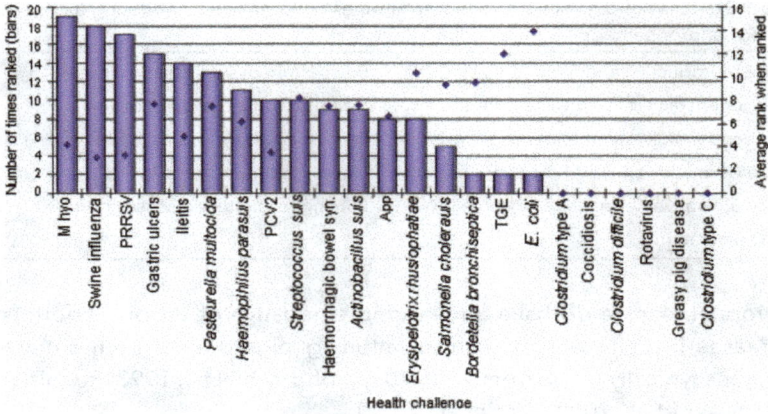

Figure 2 Rank of pathogens in the finishing herd in the United States (The most serious challenge was ranked as 1 and the other challenges were ranked in increasing order. The higher the rank, the less significant the challenge) (Holtkamp et al., 2007).

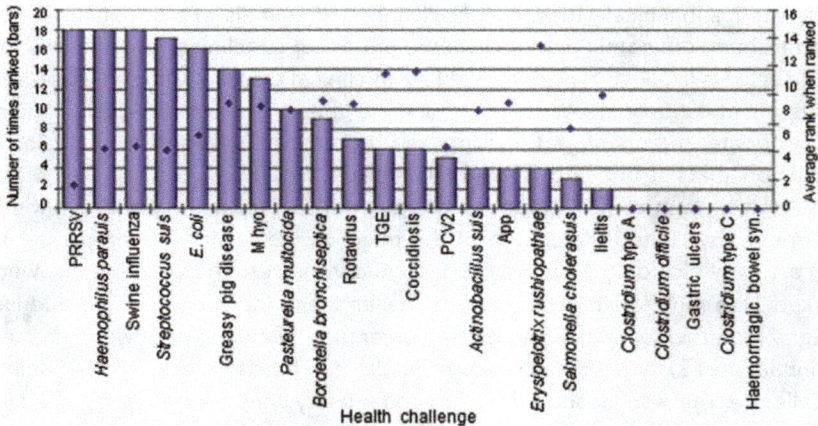

Figure 3 Rank of pathogens in the nursery herd (The most serious challenge was ranked as 1 and the other challenges were ranked in increasing order. The higher the rank, the less significant the challenge) (Holtkamp et al., 2017).

This chapter will focus on endemic pig diseases, and discuss the most important issues related to diagnosis and monitoring, and to management and control on the pig farm.

2 Disease identification

2.1 Considerations for a tentative diagnosis

For developing an efficient diagnostic approach, a clear statement of the diagnostic question should be made first (Cannon, 2002). The question might

Table 2 Summary of estimated economic losses for top four health challenges in the United States in all stages of production (Holtkamp et al., 2007)

	Losses in affected herds (USD/pig marketed)			% of animals affected			Average loss for all pigs (USD/pig marketed)			
	Breeding	Nursery	Finishing	Breeding	Nursery	Finishing	Breeding	Nursery	Finishing	Total
PRRSV	7.29	2.86	4.34	41.4	42.8	33.8	4.94	1.23	1.47	7.63
M. hyo-pneumoniae	1.52	1.92	5.84	17.6	10.0	34.3	0.39	0.19	2.00	2.58
Influenza	1.65	1.62	3.37	21.2	26.8	29.9	0.50	0.43	1.00	1.94
PRRS + M. hyopneu-moniae			6.69			18.1			1.21	1.21

be 'Why did this pig die?' or 'What is causing the clinical signs in this group of pigs?' The diagnostic approaches to address these questions may be different from those addressing questions such as 'When were these pigs exposed to the pathogen?' or 'Are these pigs still free of the pathogen?'. Enteric (e.g. diarrhoea with mortality) and systemic infections (e.g. meningitis, arthritis, with mortality) are the most important disease conditions in suckling and weaned pigs, respiratory diseases (e.g. coughing, sneezing, dyspnoea with mortality) predominate in grow-finishers and reproductive problems and lameness are most common diseases in breeding animals.

The next step in the diagnosis is to obtain a clear description of the clinical problem, including historic information. Relevant questions are: how long are the problems occurring, has the clinical picture remained the same during the last weeks and months, which age groups are affected, which are the morbidity and mortality rates, what is the response to treatment and the current vaccination status and so on. In addition to the clinical picture, information on (changes in) potential risk factors is important, for example, type of farm, purchase policy and health status of possible origin herds, management and biosecurity measures in relation to the disease, air quality, ventilation patterns and nutrition.

The historic and current farm information and the description of the problem can be easily obtained by asking proper questions to the farmer and by carefully examining the pigs and their environment. This information may allow to reach almost a final diagnosis, for example, typical outbreaks of pleuropneumonia, or else, to exclude the role of some pathogens, for example, very high mortality is unlikely to be due to single *M. hyopneumoniae* infections. In most cases, however, further diagnostic work-up is necessary to establish a conclusive diagnosis, to elucidate the pathogens involved and to assess their relative impact in the problem.

2.2 Considerations for a conclusive diagnosis

Necropsies of a representative number of typically affected animals can be very helpful. Commonly found lesions in case of respiratory disease include catarrhal pneumonia (mainly due to infections with *M. hyopneumoniae* and some viral pathogens, e.g. swine influenza), and pleurisy (mainly due to infections with *A. pleuropneumoniae, H. parasuis, S. suis, P. multocida*) (Maes et al., 2001). However, mixed infections are generally the rule and lesions may not appear very typical for one specific pathogen. Performing necropsies is mostly followed by taking samples for further diagnostic examinations, for example, histopathology, immunohistochemistry, pathogen isolation or PCR testing.

In addition to investigating dead or euthanized animals, living animals from the affected age group can be sampled to detect antibodies and/or possible pathogens involved. Different sampling strategies can be used, depending on the age group, the type of sample, the disease and the diagnostic test used (see Section 2.3.2). In case of serology, paired serum samples will often be needed to establish the diagnosis as serum antibodies appear only two to four weeks after infection. In this case, sampling during the acute phase of the disease and a second time several weeks later is required. The type and number of samples and the diagnostic tests to be used depend on the pathogen.

2.3 Monitoring health in pig herds

It may also be interesting to monitor pig herds not suffering from clinical disease in order to assess, for example, freedom of infection, infection levels and variations over time. The results may also serve as a predictor of clinical disease. Detecting disease at an early stage is very important, as treating infections in time may largely prevent or limit adverse effects on the animals' health, welfare and performance (Smart et al., 1989; Opriessnig et al., 2007; Byra et al., 2011). Different parameters can be included in a monitoring programme.

2.3.1 Clinical signs and antibiotic usage

Clinical signs in a herd can be measured on a regular basis using clinical scores, for example, a coughing or diarrhoea score (Pederson et al., 2011; Nathues et al., 2012). Scoring systems are typically based on the number of affected animals and the severity of clinical signs, and allow to describe the clinical signs in more detail and in a quantitative way. Monitoring clinical signs such as coughing can also be automated. This avoids observer bias and allows the farmer to assess the situation in the stables in real time, 24 hours a day (see further).

The registration of antibiotic usage to treat individual or groups of animals is a rather easy tool to monitor the morbidity and disease levels in a farm. Individual Pig Care is an example of a management system for monitoring the health of groups of pigs. It is based on regular, individual observation of pigs so as to permit both early detection of management and health problems and a rapid response (Pineiro et al., 2012).

2.3.2 Animal responses to infection and detection of pathogens

Different sampling strategies can be used either to assess the animal response to infection or to directly detect the pathogen. Sampling of different age groups (cross-sectional sampling and/or serial sampling) allows to investigate a large number of pigs and to assess the infection at group level. Cross-sectional sampling implies the sampling of different age groups during the same herd visit. It has the advantage that the results are obtained quickly and that only one herd visit is needed. Serial sampling implies the sampling of the same animals over time. By sampling the same group (or pigs) over time, more accurate information is obtained on the sequence of infections in these groups. However, different herd visits are needed and results are not readily available. Whenever samples are collected, it is important to use an appropriate sample size to ensure that the conclusions based on the samples are truly accurate. There are ample freeware applications available (e.g. www.winepi.net) allowing easy and straightforward sample size calculation.

To save analysis costs, individual samples may be pooled, and for some pathogens (e.g. PRRS, PCV2), oral fluids obtained from different pigs of a pen can be used (Cuevas-Cordoba and Santiago-García, 2014). It is evident that samples should be transported in a proper way (e.g. chilled vs. frozen vs. at room temperature). In addition, one should be knowledgeable about the strengths and limitations of the available diagnostic tests and the results should be interpreted properly, taking into account the predictive values of positive and negative test results, and for quantitative tests, the magnitude of the outcome.

Serology is very commonly used to assess infection levels in pig herds. It does not measure the infection itself, but the animal response to infection. Therefore, infection took place some weeks before serum antibodies are detected. Depending on the test, the type of antibodies (IgM vs. IgG) and the pathogen, the time between infection and appearance of antibodies may range from 2 (e.g. acute influenza) to 8 weeks (e.g. *M. hyopneumoniae*) (Sibila et al., 2008). Also, some infected pigs may not seroconvert, especially in case of low infection levels. As a consequence, serology should be interpreted at group level, and not at the individual level. For a proper interpretation of serological data, it is important to know whether pigs have been vaccinated, and which vaccines and vaccination schemes were used.

For many pathogens, ELISA tests are used because of the ease of use, low cost, speed of testing and possible automation. For some pathogens and/or in specific circumstances, other serological tests are used such as the haemagglutination inhibition test, for example, for swine influenza; the immunoperoxidase monolayer assay or serum neutralization test, for example, for PRRSV; and microscopic agglutination test, for example, for leptospirosis. Most serological tests do not allow discrimination between antibodies following infection or following vaccination. For some pathogens, for example, Aujeszky's disease virus, serological tests have been developed that only detect antibodies upon infection with wild virus and not following vaccination with marker vaccines.

In addition to serology, different age groups may be sampled to detect pathogens in blood (e.g. viraemia for PRRSV), tonsil samples (e.g. *A. pleuropneumoniae*), nasal samples, laryngeal or tracheal swabs or BAL fluid (e.g. *M. hyopneumoniae*) and faeces (*Salmonella, Lawsonia* (*L.*) *intracellularis, B. hyodysenteriae*). The advantage is that the pathogen itself (cultivation, electronmicrosopy), its antigens (ELISA, immunofluorescence, immunohistochemistry, *in situ* hybridization) or its genetic material (PCR) are detected, and that further characterization is possible. Successful cultivation also points to the presence of viable pathogens, and in case of bacterial pathogens, allows to perform antimicrobial sensitivity testing. Bacterial isolation is performed routinely by veterinary diagnostic laboratories and is typically considered the gold standard for unambiguous and specific detection of bacteria involved in morbidity and mortality cases. Isolation of bacterial pathogens is more difficult in case animals have received antimicrobial medication.

In general, PCR testing has a higher sensitivity compared to isolation of the pathogen. For some pathogens such as *M. hyopneumoniae*, different PCR tests are used, for example, conventional PCR, nested PCR or quantitative PCR.

2.3.3 Lesions in slaughter pigs

Respiratory tract lesions can be assessed by means of slaughter checks (Davies et al., 1995). Evaluation of respiratory tract lesions of slaughter pigs is an easy and cheap way to obtain information of the respiratory health of the fattening pigs. If the lesions are monitored from different successive batches of a herd, time evolutions can be assessed. Slaughter checks are also useful to detect subclinical respiratory infections. Most scoring methods are based on a visual subjective estimation of the proportion of lung affected surface and/or volume. Therefore, it can involve possible errors. In addition, most lung lesions are not pathognomonic for a specific pathogen. Presence of severe pleurisy may mask pneumonia lesions as well, and healing of lesions during the fattening period may lead to false-negative results (Noyes et al., 1990). Hence, the subjectivity of the scoring method, regression of lung lesions and the non-specificity of these lesions preclude an aetiologic diagnosis based solely on examination of gross lung lesions at slaughter.

Also other lesions can be monitored in slaughter pigs such as white spots on the liver due to migration of *Ascaris suum* larvae, skin lesions due to sarcoptic mange, joint lesions, gastric ulcers, tail biting lesions, abscesses, and lesions in the intestinal and reproductive tract. These lesions may also help to elucidate possible risk factors on the farm.

2.3.4 *Feed and drinking water intake*

More and more pig producers monitor feed and drinking water intake, either at barn, room or at pen level. This information is not only interesting to calculate performance, but may also be used as indirect indicators for when animals are stressed (e.g. onset of disease, excessive temperature fluctuations, heat load) or are not performing at an optimal level. The drinking behaviour of diseased pigs has been found to deviate from the consistent patterns seen in healthy pigs (Maselyne et al., 2015). The behavioural changes appear in the subclinical stage of infection, before disease symptoms become visually apparent. If water usage drops more than 30% in a day, or drops for three consecutive days, a potential health challenge may occur.

3 Disease management and control: overview

3.1 *Types of disease prevention*

Disease prevention may be considered at different levels (Toma et al., 1999) (Table 3). Primary prevention relates to measures that prevent the introduction of pathogens and/or of related determinant factors in a herd. It is the most effective form of prevention. If the pathogen is not present, then disease caused by the pathogen is not possible. Many different strategies to eliminate pathogens from pig herds have been described (Harris and Alexander, 1999; Holst et al., 2015). Eradication programmes for specific diseases, for example, classical swine fever (CSF) and Aujeszky's disease have been successful in the EU. Also endemic infections such as PRRSV, atrophic rhinitis, *B. hyodysenteriae*

Table 3 Types of disease prevention in pig herds (adapted from Toma et al., 1999)

	Type of prevention		
	Primary	*Secondary*	*Tertiary*
Goal	Absence of pathogen	Absence of disease	Effective treatment of disease
Strategy	Control of risk factors and causes	Control of infection and disease	
Method	Health management programmes (especially external biosecurity)	Health management programmes (external and internal biosecurity, management, nutrition, etc.)	Therapeutic measures

and *M. hyopneumoniae* have been eliminated successfully from individual herds. Although elimination is the most straightforward measure, there is a permanent risk for re-infections. This is especially the case for herds located in regions with a high pig (herd) density and for pathogens that can spread via airborne transmission, for example, *M. hyopneumoniae* (Maes et al., 2008). To account for these risks, air filtration systems have been installed in modern facilities (see further).

In countries with intensive pig production systems, most herds are endemically infected with several (facultative pathogenic) pathogens. For a number of reasons, it is very difficult and/or not warranted to eliminate all of these pathogens from these herds. Secondary prevention implies that pathogens are still present on the herd, but the infection level is in balance with the herd immunity thanks to optimal management and biosecurity. Finally, clinically diseased pigs should be treated properly as quickly as possible. According to Toma et al. (1999), this is called tertiary prevention. Although this can hardly be considered as prevention, it will remain important in the future. In general terms and especially in farm animals, prevention is better than cure. Therefore, emphasis should be placed on control and prevention of infections.

3.2 Biosecurity measures and management: definition and importance

Biosecurity is the application of management practices that reduce the opportunities for infectious agents to gain access to, or spread within, a food animal production unit (Toma et al., 1999). External biosecurity comprises measures that prevent pathogens from entering the herd, while internal biosecurity relates to preventing the within-herd spread of pathogens. Biosecurity encompasses cleanliness; disinfection; reduction of exposure; management of animals, personnel and goods; and ensuring that animals can be traced. Biosecurity is therefore, essentially, the application of 'informed common sense'. It is an everyday challenge for the farmer. Recent studies have quantified biosecurity on pig herds and the score was expressed as a percentage (Laanen et al., 2013) with 100% being the perfect biosecurity situation and 0% the total absence of any biosecurity measures (www.biocheck.ugent.be). The average external biosecurity score was 65 (range: 45-89) and the average internal score was 52 (range: 18-87). External scores were positively associated with herd size, while internal scores were negatively associated with both 'age of buildings' and 'years of experience of the farmer', indicating that biosecurity is generally better implemented in larger herds, in more modern facilities and by younger farmers. External and internal biosecurity scores were positively associated with performance of fattening pigs. Internal scores were negatively associated with disease treatment incidence, suggesting that improved biosecurity might help in reducing the prophylactic use of antimicrobials. The explained variation in the outcome variables, however, were low, indicating that other factors apart from the biosecurity parameters are also important. Similar results have been obtained in more recent studies (Postma et al., 2016). Biosecurity scoring is particularly interesting for sensitizing pig producers and benchmarking of herds. The most important factors related to external and internal biosecurity are discussed below.

4 External biosecurity

4.1 Purchase policy

Transmission of pathogens occurs very effectively via direct contact between infected and susceptible animals. Tobias et al. (2014) described a ten times higher transmission rate of A. pleuropneumoniae for direct contact between pigs compared to indirect transmission. Similar findings were obtained for other swine pathogens. Therefore, purchasing animals from other pig herds constitutes a risk for pathogen introduction, especially when it is performed at a regular basis, when replacement rates are high and in case animals originating from different sources. When it is necessary to purchase animals (e.g. to improve genetics), it is important to evaluate the health status of the origin herd and to limit the number of sources. Moreover, if new animals are introduced in a herd, adequate quarantine measures should be taken (Amass and Baysinger, 2006; Maes et al., 2008).

According to Calvar et al. (2014), pig farmers in Bretagne are not fully aware of the risk of purchasing breeding stock in pig farms in Bretagne, except for PRRSV. In an earlier study on quarantine (Calvar et al., 2012), they observed that barely 60% of the breeders know the health status of the delivered animals. On the other hand, the majority of Danish farms have adopted the specific-pathogen-free (SPF) concept and follow a much stricter protocol. To date, approximately 3100 herds in Denmark have a SPF health declaration, while 78% of the sows in Denmark are SPF sows. The SPF system is built on the principle that health status is declared for the connected herds and is taken into account in case of trading between herds. It combines documented high biosecurity at herd level, regular disease testing and publishing of disease status, among other things, according to the certified disease status of the herds (Anonymous, 2015). A previous study on biosecurity in Belgian pig farms showed a good level of implementation of the measures related to purchase policy of animals and semen (Laanen et al., 2013).

Purchasing contaminated semen is also known to be a risk. Pathogen transmission by semen to the sow has been proven for viruses such as CSF virus and PRRSV (Maes et al., 2016). The best way to prevent disease transmission via the semen is to ensure that the boars in artificial insemination centres are free from the pathogen.

4.2 Dead animals and manure

Secreta and excreta from diseased and dead animals are often contaminated with pathogens and pose a risk for direct transmission to susceptible animals. Also indirect transmission of pathogens from cadavers may occur, for example, via fomites, the rendering truck, people (including the veterinarian) and their material, rodents and domestic animals. The risk of transmission via infectious cadavers is well known for foot-and-mouth disease (Scudamore et al., 2002; Hayama et al., 2015). A timely, consequent and careful removal of dead animals is therefore recommended. Many pig pathogens, for example, *Brachyspira* spp. (Hampson, 2012) or *Salmonella enterica* (Carlson et al., 2012) are able to survive for long periods in manure. Therefore, manure or utensils contaminated with manure constitute a risk for introduction of pathogens.

4.3 Feed, water and equipment

Trucks delivering commercial feed to farms enter multiple premises a day often without being (properly) cleaned and disinfected in-between. This causes a risk for pathogen transmission between farms as described by Bottoms et al. (2015). One explanation for the transmission of disease in the recent outbreak of porcine epidemic diarrhoea (PED) virus in North America was the use of contaminated bulk feed containers (Scott et al., 2016). Feed itself should not normally pose a risk of disease introduction since it is generally produced under strict hygienic procedures, although introduction of *Salmonella enterica* via feed has been described (Österberg et al., 2006). Swill feeding, however, causes substantial risks as demonstrated in several CSF outbreaks in the past. Swill feeding is currently banned in the EU (European Commission, 2001). The risk for disease introduction via the feed is much lower for farmers who home mix feed.

Water might also play a role in disease transmission since pathogens may survive in water (Loera-Muro et al., 2013; Szabo and Minamyer, 2014). Duration of survival depends on different factors including traits of both the pathogen and the drinking water. Therefore, a regular (at least once a year) microbiological and chemical evaluation of the water quality is advisable.

4.4 Access check

The farmer, employees, the veterinarian and health professionals, or herd visitors can function as mechanical vectors by spreading pathogens via, for example, their footwear or clothes, but also via their hands. Ribbens et al. (2007) showed that contaminated boots, gloves and coveralls may be involved in the transmission of CSF virus. Amass et al. (2003) showed that showering was able to prevent transmission of *S. suis* from a person to the pigs.

4.5 *Vermin and bird control*

Rodents and especially rats are well known to spread pathogens (Pearson et al., 2016). Pathogens known to be transmitted by rodents include *L. intracellularis, B. hyodysenteriae* and *Salmonella enterica*. The same authors also showed that birds may transmit pathogenic *E. coli* to pigs. Domestic animals such as dogs and cats may also introduce or transmit pathogens. Desrosiers (2011) reported that not only dogs but also insects were able to transmit *B. hyodysenteriae*, and that cats and dogs could transmit *P. multocida*. Saif et al. (2012) reported the shedding of transmissible gastroenteritis as well as PED virus by dogs and cats, and Opriessnig and Wood (2012) described domestic animals as potential reservoirs for *Erysipelothrix rhusiopathiae*.

4.6 *Herd location and environment*

Some pathogens, for example, *M. hyopneumoniae*, swine influenza virus and PRRSV can be transmitted between herds via airborne transmission (Otake et al., 2010; Desrosiers, 2011). Almost all pathogens, especially respiratory pathogens, can be found in the air in the close proximity of the animals (Tobias et al., 2014). The location of the herd and the density of pig farms in the proximity are thus important risk factors. Maes et al. (2000) reported increased seroprevalences for *M. hyopneumoniae*, swine influenza viruses and Aujeszky's disease virus in herds located in pig-dense areas. Filtration of incoming air, in combination with standard biosecurity procedures, has been demonstrated to prevent transmission of PRRSV into susceptible herds (Alonso et al., 2013). The authors showed that air filtration reduced the risk of introduction of novel PRRSV by approximately 80%, indicating that on large sow farms with good biosecurity in pig-dense regions, approximately four out of five PRRSV outbreaks may be attributable to aerosol transmission. Air filtration may also be helpful to prevent airborne transmission of other pathogens such as swine influenza and *M. hyopneumoniae*.

Feral animals, with an important role of wild boar, are also able to transmit several pig pathogens to domestic pigs.

5 Internal biosecurity

5.1 *Production system*

All-in, all-out (AIAO) production is an important factor in the control of infectious disease since it can interrupt or decrease the cycle of pathogen transmissions from older to younger pigs. It allows the producer to tailor environmental conditions to a uniform population of pigs and to clean the facilities between groups of pigs. AIAO production also results in better performance and less lung lesions in slaughter pigs (Straw, 1991). Mixing or sorting pigs is a source of stress to the animals and it increases the probabilities of disease transmission. Therefore, an AIAO system in which the same pigs are moved as a group through the different production stages is to be preferred compared to one where pigs are regrouped during transfer from one unit to another.

Early weaning (<3 weeks) can reduce transmission of some pathogens from the sow to the offspring, but it is not allowed to be applied systematically in the EU. In addition, recent studies have shown that early weaning may increase the antimicrobial consumption post-weaning (Postma et al., 2016). Parity segregation has been used in large production systems as a means to control several diseases in the breeding herd. Gilts and their offspring are kept separately from the sows until they reach their second gestation. By that time, they are expected to have acquired the desired immune status and pose no risk for destabilizing the infection level in the herd (Hoy et al., 1986).

5.2 Compartmentalizing, working lines and equipment

Keeping age groups separately and working in a well-defined sequence is recommended, that is, first the youngest animals, followed by the older age groups, thereafter the quarantine and sick bay, and finally the cadaver storage (Neumann, 2012).

Materials and equipment are very important in the transmission of a large number of pathogens. Allaart et al. (2013) described the isolation of *Clostridium perfringens* from bedding material, drinking water, boots, fans and fly strips on farms. Syringes and needles might also serve as transmitters of disease, for example, for PRRSV (Otake et al., 2002) and PCV2 (Patterson et al., 2011).

5.3 Animal stocking density

Decreasing animal density during the different production stages has been shown to reduce the level of disease (Pointon et al., 1985). Crowding may lead to increased transmission of pathogens and stress reactions, making the pigs more susceptible to infectious diseases. Too low stocking densities, however, are financially not justified. Therefore, it is important to find a reasonable compromise between stocking densities that are appropriate for the health of the pigs and those that maximize the returns on the building's cost.

The stocking densities based on EU legislation (Table 4) have been determined in the 1980s, and should not necessarily be considered as optimal values (Dewulf et al., 2007). Studies suggest that for optimal health and production, surface areas per pig in the different age categories may be up to 20% higher (more surface area per pig) than the minimal legal requirements

Table 4 Stocking densities for pigs according to EU legislation (Directive 2008/120/EC)

Average weight (kg)	Minimal required surface area (in m²) per pig
<10	0.15
10 to 20	0.20
20 to 30	0.30
30 to 50	0.40
50 to 85	0.55
85 to 110	0.65
>110	1.00

(Hamilton et al., 2003). Further research on the influence of different stocking densities on health, welfare and performance is needed, including also economic considerations.

5.4 Management of diseases

Diseased animals may pose a substantial threat to susceptible animals. Correct handling and treatment of diseased animals is therefore of great importance in the reduction of transmission of pathogens.

5.5 Herd size

Study results on the role of herd size as a risk factor for disease are not consistent. Flesja and Solberg (1981) found an increased prevalence of *M. hyopneumoniae* lesions in larger pig herds, whereas in other studies no significant associations could be found between herd size and seroprevalence of *M. hyopneumoniae* (Maes et al., 1999, 2000) or prevalence of *Mycoplasma*-like lung lesions at slaughter (Maes et al., 2001). Biologically plausible reasons for a positive association between herd size and respiratory infections include a greater risk of introduction of the pathogen from outside the herd, greater risk of pathogen transmission within and among herds when the herd is large, and effects of management and environmental factors that are related to herd size. However, owners of large herds might more frequently and more rapidly adopt management and housing practices that mitigate this potential increased risk (Gardner et al., 2002; Laanen et al., 2013).

5.6 Improvement of housing conditions

Housing and/or environmental changes that optimize the climate of the pigs' environment are important in the control of infections. In case of respiratory disease, special attention should be paid to the temperature set points, fan staging, air inlet and curtain settings, sensor placement, heater capacity, drafts and building maintenance (Gonyou et al., 2006). However, making environmental changes for improving the climate in inappropriate or old barns frequently entail extensive remodelling, and therefore, they may be difficult and expensive to implement. Worn slats are more difficult to clean and disinfect, and constitute a risk for enteric pathogens.

5.7 Cleaning and disinfection

Following the correct procedure of cleaning, disinfection and empty period in the compartments/pens will reduce the infection pressure. A lack of cleaning after batches was seen as a risk factor for several zoonotic pathogens such as *Salmonella enterica*, *Campylobacter* spp. and *Listeria monocytogenes* (Fosse et al., 2009; De Busser et al., 2013). Hygiene testing, that is, testing of bacterial contamination of surfaces, or adenosine triphosphate (ATP) analyses of surfaces can be used to check whether the cleaning and disinfection were performed properly.

5.8 Nutrition and drinking water

Nutrition is a key factor for pig health. The feed composition and the physical characteristics of the feed, the feeding level as well as the way of feeding and feeder space per pig are all important. The severity of endemic diseases such as infections with *E. coli* post-weaning, *Brachyspira* spp. and *Salmonella enterica* are largely influenced by nutritional aspects. Feeding coarsely ground meal decreased the survival of *Salmonella enterica* during passage through the stomach (Mikkelsen et al., 2004). A strong reducing effect of fermented liquid feed, including whey, on *Salmonella enterica* shedding and seroprevalence has also been reported (Poljak et al., 2008). The addition of organic acids to feed or drinking water is generally beneficial to reduce *E. coli* and *Salmonella enterica* shedding, but the effects are variable (De Busser et al., 2009; De Ridder et al., 2013).

6 Vaccination and antimicrobial medication

6.1 Vaccination

Improving the immunity status of susceptible animals, for example, by vaccination is a helpful tool to control infectious diseases. Vaccination is generally able to reduce the (risk for) clinical symptoms, lesions and performance losses due to disease. However, most vaccines only provide partial protection, do not prevent infection and are not able to eliminate the pathogens from the herd. The decision whether to implement vaccination is not straightforward. It depends on the pig producer (degree of being 'risk averse'); the type and severity of disease; the available vaccines and expected effects; and other factors such as type of herd, flow of the pigs, infection level, management and housing conditions. The herd veterinarian has a central role for determining which pathogens are playing a key role, and for making a decision in collaboration with the farmer to implement one or more vaccinations and determining the optimal vaccination scheme.

Vaccination schemes are highly variable between herds and also between countries. Some pathogens may be eradicated in some countries and vaccination is prohibited (e.g. Aujeszky's disease virus in many European countries and North America), whereas the virus is still endemic in many other countries and vaccination is used.

Breeding sows are often vaccinated against PRRSV, neonatal *E. coli* infections, parvovirus and erysipelas, and in some herds also against other pathogens such as atrophic rhinitis, swine influenza, PCV-2, *H. parasuis*, *A. pleuropneumoniae*, leptospirosis and so on. In Asian countries, farms may also vaccinate against foot-and-mouth disease, Aujeszky's disease, CSF and Japanese encephalitis. Some vaccinations in sows aim to provide protection to the piglets via maternally derived antibodies, for example, neonatal *E. coli* infections and atrophic rhinitis. Recent studies in high-performing sow herds have shown that colostrum production is quite variable between herds and between sows, and that for many piglets, colostrum intake is insufficient (Decaluwé et al., 2014; Declerck et al., 2015).

Commonly used commercial vaccines in pigs include vaccines against infections with PCV-2, *M. hyopneumoniae*, PRRSV and depending on the country and the herd also vaccines against other pathogens may be applied such as *A. pleuropneumoniae*, swine influenza, *H. parasuis*, post-weaning diarrhoea, oedema disease, Aujeszky's diseases virus and so on.

The costs and benefits of disease control campaigns can be assessed using several methods including gross margin analysis and partial budgeting (Thrusfield, 2007). These are essentially straightforward accounting approaches. Social cost–benefit analysis is the application of a specific technique that allows for the fact that costs and benefits are commonly distributed over time, and sometimes are more than simple financial values.

Maes et al. (2003) conducted a multi-site field study to assess the efficacy and economic profitability of using an inactivated *M. hyopneumoniae* vaccine in 14 pig herds infected by *M. hyopneumoniae* and practising an AIAO production system. Vaccination significantly improved daily weight gain (+22 g), feed conversion ratio (−0.07), medication costs (−0.476 €/pig), prevalence of pneumonia lesions (−14%) and severity of pneumonia lesions (−3%). Mortality rate, severity of coughing and carcass quality were not significantly influenced by vaccination. A partial budget analysis based on significantly improved performance parameters demonstrated that *M. hyopneumoniae* vaccination was economically attractive as it resulted in an increase of the net return to labour with €1.3 per finishing pig sold. The sensitivity of the economic benefit was illustrated towards fluctuations in pig finishing prices.

In general, vaccination is economically justified in clinically affected herds, especially if no major risk factors can be identified or changed in short term. However, vaccination may also be cost-efficient in subclinically infected herds as subclinical infections, for example, with *M. hyopneumoniae*, *A. pleuropneumoniae*, swine influenza, PRRSV and PCV2 also reduce performance (Rohrbach et al., 1993; Regula et al., 2000; Maes et al., 2003). For some important pig pathogens, however, for example, *B. hyodysenteriae* and *S. suis*, no or few commercial vaccines are available. Autogenous vaccines may also be used. These are mostly inactivated vaccines, for example, bacterins based on the pathogen strain(s) isolated from animals affected by the disease. Autogenous vaccines against different pathogens such as *S. suis*, *A. pleuropneumoniae* and *E. coli* have been described (Baums et al., 2010; Geldhof et al., 2012). One of the disadvantages of autogenous vaccines is that often limited data are available on vaccine efficacy and safety (Haesebrouck et al., 2004).

6.2 Antimicrobial medication

Antimicrobials are necessary for treatment of bacterial infections and by doing so to safeguard animal welfare. They should, however, be used judiciously and restrictively to avoid or limit the risk for antimicrobial resistance in both pathogens and the commensal flora. High antimicrobial use and the resulting risk for antimicrobial resistance is an important concern worldwide. Research showed that prophylactic use of antimicrobials is common in many pig herds (Callens et al., 2012). At the same time, variations exist in the average levels of treatment between countries (Sjölund et al., 2016). These variations are beyond the difference in general level of pig health and therefore reflect also habits and attitudes in relation to antimicrobial use. Repeated prophylactic use without sufficient diagnostic data or using antimicrobials as a substitute for improper management cannot be justified. In addition, it can be expected that the use of critically important antimicrobials such as fluoroquinolones and third- and fourth-generation cephalosporins will be further restricted or (in some countries) even be prohibited for use in veterinary medicine. A comparison between the use of vaccination and antimicrobial medication for the treatment and control of bacterial disease in pigs is presented in Table 5.

Table 5 Advantages and disadvantages of vaccination versus antimicrobial medication to control disease in pig herds

Vaccination	Antimicrobial medication
Long-term strategy	Short-term, flexible
More labour-intensive	Less labour-intensive (oral medication)
Against one pathogen (combination vaccines possible)	Against different pathogens (e.g. multiple disease challenges)
Effect until end of fattening period	Effects mainly during medication period
No risk of residues in carcasses	Risk of residues in carcasses
No selection of antibiotic resistance	Selection of antibiotic resistance
Vaccines are available against a limited number of diseases	Against bacterial infections, not against viruses
Partial protection	Partial protection

7 Future trends in diagnostics and disease monitoring and control

7.1 Precision livestock farming

Precision livestock farming (PLF) has the potential to revolutionize livestock farming and help solve many of the challenges mentioned in the previous sections (Banhazi et al., 2012). It implies the automated (real-time) monitoring or problem detection in livestock. Measurements at the individual level capture the large variation between pigs, and may also be interesting for welfare and health parameters (Berckmans, 2006). Examples include the measurement of feeding and drinking behaviour, body weight, activity, body temperature or other clinical signs such as cough (Cornou and Kristensen, 2013). Maselyne et al. (2015) showed that a radio-frequency identification system for measuring drinking behaviour of fattening pigs could be used as an on-farm detection system for health, welfare and production problems. Infrared temperature measurement equipment is gaining popularity as a diagnostic tool for evaluating human and animal health (Soerensen and Pedersen, 2015).

Systems based on group-level measurements are, however, generally cheaper and easier to implement than systems based on individual measurements. Berckmans et al. (2015) showed that the respiratory distress monitor, a tool that automatically monitors the respiratory health status of a group of pigs, gives earlier warnings compared to a situation where the farmer and veterinarian rely on their own routine observations without the monitor. Sound-based PLF techniques have advantages over other technologies such as cameras or accelerometers because microphones are contactless and relatively cheap, there is no need for a direct line of sight and large groups of animals can be monitored with a single sensor in a room.

Finally, irrespective of the system or the parameter(s) measured, PLF does not aim to replace the farmer but should be considered as a tool to support management decisions.

7.2 *Diagnostics*

The rapid and accurate establishment of a microbial cause is fundamental to quality porcine health management. Despite dramatic advances in diagnostic technologies, many groups of pigs with suspected infections receive empiric antimicrobial therapy rather than appropriate therapy dictated by the rapid identification of the infectious agent. New tests are needed that can identify one or preferably more specific pathogens, or at a minimum, distinguish between bacterial and viral infections. Also, they should provide information on susceptibility to antimicrobial agents. Tests should be easy to use, at the farm (pen side tests), and provide a rapid, but reliable, result (ideally within an hour).

The expertise of the herd veterinarian to interpret the results will become more important with the advent of newer, more complex tests. The availability of needed tests will lead to improvements in clinical outcomes upon treatment, antimicrobial stewardship, detection and tracking of disease outbreaks, and investigation of unknown pathogens. Emerging technologies could enable the detection and quantification of pathogen burden with new speed, sensitivity and simplicity of use. However, there are significant challenges to the development, regulatory approval, and last-but-not-least, the cost of diagnostic tests based on these new technologies.

7.3 *Control measures*

Further research into disease prevention and infection control measures such as biosecurity and hygiene procedures are warranted. Many of the current practices are based on empirical expertise, whereas more in-depth research could provide a better insight in and development of preventive measures. In addition, research into the importance of the microbiota for pig health and production is needed.

More and also better vaccines will need to be developed in the future, as for some important pig diseases (e.g. *Brachyspira* spp., *S. suis*) no or very few commercial vaccines are available. Many currently used commercial vaccines only provide partial protection. Therefore, vaccines that confer a better protection are needed. To achieve this, more research is warranted to understand the pathogenesis of the disease, and the protective mechanisms upon infection. It can also be expected that autogenous vaccines, for example, vaccines made to suite a specific herd health situation, may be more frequently used in the future, especially if commercial vaccines are not available or do not provide sufficient protection. Research should also focus on easy administration routes for vaccines, eventually by oral or aerosol administration, and on developing multivalent vaccines. A better understanding of the composition and the role of the pig microbiota is needed, as this may help to better understand the mode of action of probiotics and prebiotics, and may enhance the development of new products to control enteric disease.

7.4 Selection for disease resistance

Many diseases of both farm and companion animals have a variable heritable component. Resistance can be defined as the ability of the host to exert some degree of control over the pathogen life cycle (Bishop, 2012). It encompasses the many ways a host species may be more resistant, for example, less likely to become infected, reduced pathogen proliferation once infected, reduced shedding or transmission of infection. Genetic and genomic studies of disease resistance have been conducted, elucidating the genetic control of between-animal differences in resistance (Bishop and Woolliams, 2014). Notable examples of success include infectious pancreatic necrosis in Atlantic salmon and bovine tuberculosis in dairy cattle. In pigs, Whitworth et al. (2015) edited the gene that makes the CD163 protein which is important during infection with PRRSV. Treated pigs did no longer produce the protein and were protected against clinical disease upon experimental PRRS challenge infection. The results of these studies may be utilized to breed animals for increased resistance. The development of genetic resistance against diseases in pigs has, however, been slow and uneven so far. It has been a long and ultimately unsuccessful battle to develop useful *E. coli*-resistant piglets.

The use of genetic or genomic information must be considered in the broader context of both the breeding goal and the conventional strategies used to control the disease. If the disease is of little importance to the breeder or producer, or is satisfactorily controlled by other means, then there may be little point in attempting to breed for increased resistance. Consequently, most effort should be directed towards costly endemic diseases, for which control by other strategies is proving difficult.

8 Conclusion

Pig health is essential for a sustainable pig production, as infectious diseases may cause substantial economic losses and may affect the quality and safety of pork. As most pathogens in pig herds are present in a subclinical form, monitoring is important to assess the infection levels and to identify the major pathogens involved. Animals in intensive production systems are continuously in a fragile balance. Optimal management, biosecurity and nutrition are needed to maintain health and production targets. Vaccination and medication may be needed as additional control measures. Further research is needed to better understand the host–pathogen interactions and to develop better diagnostics, vaccines and control measures against infectious pig diseases.

9 Where to look for further information

Diseases of Swine. Zimmermann, J., Karriker, L., Ramirez, A., Schwartz, K. and Stevenson, G. (eds), Wiley-Blackwell, Chichester, West Sussex, UK.

Porcine Health Management. The official journal of the European College of Porcine Health Management and the European Association of Porcine Health Management: https://porcinehealthmanagement.biomedcentral.com/.

Journal of Swine Health and Production. The official journal of the American Association of Swine Veterinarians: https://www.aasv.org/shap.html.

Unit of Porcine Health Management, Department of Reproduction, Obstetrics and Herd Health, Faculty of Veterinary Medicine, Ghent University, Belgium: http://www.rohh. ugent.be/v3/research/units/pig_health/.

10 References

Allaart, J., van Asten, A. and Gröne A., 2013. Predisposing factors and prevention of *Clostridium perfringens*-associated enteritis. Comp. *Immunol. Microbiol. Infect. Dis.* 36, 449-64.

Allan, G. and Ellis, J., 2000. Porcine circoviruses: a review. *J. Vet. Diagn. Invest.* 12, 3-14.

Alonso, C., Murtaugh, M., Dee, S. and Davies, P., 2013. Epidemiological study of air filtration systems for preventing PRRSV infection in large sow herds. *Prev. Vet. Med.* 112, 109-17.

Alvarez-Ordóñez, A., Martinez-Lobo, F., Arguello, H., Carvajal, A. and Rubio, P., 2013. Swine dysentery: aetiology, pathogenicity, determinants of transmission and the fight against the disease. *Int. J. Environ. Res. Public Health* 10, 1927-47. https://doi. org/10.3390/ijerph10051927

Amass, S. and Baysinger, A., 2006. Swine disease transmission and prevention. In *Diseases of Swine*, 9th edn, Straw, B., Zimmerman, J., D'Allaire, S. and Taylor, D. (eds), Chapter 68, 1075-98.

Amass, S., Halbur, P., Byrne, B., Schneider, J., Koons, C., Cornick, N. and Ragland, D., 2003. Mechanical transmission of enterotoxigenic *Escherichia coli* to weaned pigs by people, and biosecurity procedures that prevented such transmission. *J. Swine Hlth. Prod.* 11, 61-8.

Anonymous, 2015: SPFSUS. SPF-Sundhedsstyringen – Status June 2015. Available at: http://spfsus.dk/en (accessed 18 November 2016).

Banhazi, T., Lehr, H., Black, J., Crabtree, H., Schofield, P., Tscharke, M. and Berckmans, D., 2012. Precision livestock farming: An international review of scientific and commercial aspects. *IJABE* 5, 1-9.

Baums, C., Brüggemann, C., Kock, C., Beineke, A., Waldmann, K. and Valentin-Weigand, P., 2010. Immunogenicity of an autogenous *Streptococcus suis* bacterin in preparturient sows and their piglets in relation to protection after weaning. *Clin. Vaccine Immunol.* 10, 1589-97.

Berckmans, D. 2006. Automatic on-line monitoring of animals by precision livestock farming. In *Livestock Production and Society*, Geers, R. and Madec, F. (eds), International Society for Animal Hygiene, Wageningen Academic Publishers, Wageningen, The Netherlands, 51-4.

Berckmans, D., Hemeryck, M., Berckmans, D., Vranken, E. and van Waterschoot T., 2015. Animal sound … talks! real-time sound analysis for health monitoring in livestock. *International Symposium on Animal Environment & Welfare*, 23-26 October, 2015, Chongqing, China, 1-8.

Bishop, S., 2012. A consideration of resistance and tolerance for ruminant nematode infections. *Front Genet.* 3, 168. https://doi.org/10.3389/fgene.2012.00168

Bishop, S. and Woolliams, J., 2014. Genomics and disease resistance studies in livestock. *Livest. Sci.* 166, 190-8.

Bottoms, K., Dewey, C., Richardso, K. and Poljak, Z., 2015. Investigation of biosecurity risks associated with the feed delivery: a pilot study. *Can. Vet. J.* 56, 502-8.

Byra, F., Gadbois, P., Cox, W., Gottschalk, M., Farzan, V., Bauer, S. and Wilson J., 2011. Decreased mortality of weaned pigs with *Streptococcus suis* with the use of in-water potassium penicillin G. *Can. Vet. J.* 52, 272-6.

Callens, B., Persoons, D., Maes, D., Laanen, M., Postma, M., Boyen, F., Haesebrouck, F., Butaye, P., Catry, B. and Dewulf, J., 2012. Prophylactic and metaphylactic antimicrobial use in Belgian fattening pig herds. *Prev. Vet. Med.* 106, 53-62.

Calvar, C., Heugebaert, S., Caille, M. E. and Roy, H., 2012. La quarantaine. Des préconisations de techniciens diversifies. Des conduits multiples chez de très bons éleveurs. Rapport d'étude, Chambres d'agriculture de Bretagne.

Calvar, C. and Lemoine, T., 2014. La biosécurité en élevage de production. Rapport d'étude, Chambres d'agriculture de Bretagne.

Cannon, R. 2002. Demonstrating disease freedom-combining confidence levels. *Prev. Vet. Med.* 52, 227-49.

Carlson, S., Barnhill, A. and Griffith, R., 2012. Salmonellosis. In *Diseases of Swine*, Zimmermann, J. J., Karriker, L. A., Ramirez, A., Schwartz, K. J. and Stevenson, G. W. (eds), Wiley-Blackwell, Chichester, West Sussex, UK, 821-33.

Cornou, C. and Kristensen, A., 2013. Use of information from monitoring and decision support systems in pig production: collection, applications and expected benefits. *Livest. Sci.* 157, 552-67.

Cuevas-Cordoba, B. and Santiago-Garcıa, J., 2014. Saliva: a fluid of study for OMICS. *OMICS* 18, 87-97.

Davies, P., Bahnson, P., Grass, J., Marsh, W. and Dial, G., 1995. Comparison of methods for measurement of enzootic pneumonia lesions in pigs. *Am. J. Vet. Res.* 56, 709-14.

De Busser, E., Dewulf, J., Nollet, N., Houf, K., Schwarzer, K., De Sadeleer, L., De Zutter, L. and Maes D., 2009. Effect of organic acids in drinking water during the last 2 weeks prior to slaughter on *Salmonella* shedding by slaughter pigs and contamination of carcasses. *Zoonoses Public Health* 56, 129-36.

Decaluwé, R., Maes, D., Wuyts, B., Cools, A. and Janssens, G., 2014. Piglets' colostrum intake is associates with daily weight gain and survival until weaning. *Livest. Sci.* 162, 185-92.

Declerck, I., Dewulf, J., Piepers, S., Decaluwé, R. and Maes, D., 2015. Sow and litter factors influencing colostrum yield and nutritional composition. *J. Anim. Sci.* 93, 1309-17.

De Ridder, L., Maes, D., Dewulf, J., Pasmans, F., Boyen, F., Méroc, E., Butaye, P. and Van der Stede, Y., 2013. Evaluation of three intervention strategies to reduce the transmission of *Salmonella typhimurium* in pigs. *Vet. J.* 197, 613-8.

Desrosiers, R., 2011. Transmission of swine pathogens: different means, different needs. *Anim. Health Res. Rev.* 12, 1-13.

Dewulf, J., Tuyttens, F., Lauwers, L., Van Huylenbroeck, G. and Maes, D., 2007. De invloed van de hokbezettingsdichtheid bij vleesvarkens op productie, gezondheid en welzijn. *Vl Diergeneesk Tijdschr* 76, 410-6.

Díaz, I., Cortey, M., Darwich, L., Sibila, M., Mateu, E. and Segalés, J., 2012. Subclinical porcine circovirus type 2 infection does not modulate the immune response to an Aujeszky's disease vaccine. *Vet. J.* 194, 84-8. https://doi.org/10.1016/j.tvjl.2012.02.014

European Commission, 2001. Council Directive 2001/89/EC of 23 October 2001 on Community measures for the control of classical swine fever. In European Commission (Ed.), 2001/89/EC. Official Journal of the European Communities, Brussels, Belgium.

Flesja, K. and Solberg, I., 1981. Pathological lesions in swine at slaughter. *Acta Vet. Scand.* 22, 272-82.

Fosse, J., Seegers, H. and Magras, C., 2009. Prevalence and risk factors for bacterial food-borne zoonotic hazards in slaughter pigs: a review. *Zoonoses Public Health* 56, 429-54.

Gardner, A., Willeberg, P. and Mousing, J., 2002. Empirical and theoretical evidence for herd size as a risk factor for swine diseases. *Anim. Health Res. Rev.* 3, 43-55.

Geldhof, M., Vanhee, M., Van Breedam, W., Van Doorsselaere, J., Karniychuk, U. and Nauwynck, H., 2012. Comparison of the efficacy of autogenous inactivated Porcine Reproductive and Respiratory Syndrome Virus (PRRSV) vaccines with that of commercial vaccines against homologous and heterologous challenges. *BMC Vet. Res.* 8, 182.

Gonyou, H., Lemay, S. and Zhang, Y., 2006. Effects of the environment on productivity and disease. In *Diseases of Swine*. 9th edn. Straw, B., Zimmerman, J., D'Allaire, S. and Taylor, D. (eds), Blackwell Publishing Ltd., Oxford, UK, 1027-38.

Haesebrouck, F., Pasmans, F., Chiers, K., Maes, D., Ducatelle, R. and Decostere, A., 2004. Efficacy of vaccines against bacterial diseases in swine: what can we expect? *Vet. Microbiol.* 100, 255-68.

Hamilton, D., Ellis, M., Wolters, B., Schinckel, A. and Wilson, R., 2003. The growth performance of the progeny of two swine sire lines reared under different floor space allowances. *J. Anim. Sci.* 81, 1126-35.

Hampson, D., 2012. Brachyspiral colitis. In *Diseases of Swine*. Zimmermann, J., Karriker, L., Ramirez, A., Schwartz, K. and Stevenson, G. (eds), Wiley-Blackwell, Chichester, West Sussex, UK, 680-96.

Harris, D. and Alexander, T., 1999. Methods of disease control In Straw, B., D'Allaire, S., Mengeling, W. and Taylor, D. (eds), *Diseases of Swine*. Iowa State University Press, Ames, US, 1077-110.

Hayama, Y., Kimura, Y., Yamamoto, T., Kobayashi, S. and Tsutsu, T., 2015. Potential risk associated with animal culling and disposal during the foot-and-mouth disease epidemic in Japan in 2010. *Res. Vet. Sci.* 102, 228-30.

Holtkamp, D., Rotto, H. and Garcia, R., 2007. The economic cost of major health challenges in large US swine production systems. In Proc. AASV, 3-6 March 2007, Orlando, 85-9.

Hoy, S., Mehlhorn, G., Hörügel, K., Dorn, W., Eulenberger, K. and Johannsen, U., 1986. Der Einfluss ausgewählter endogener Faktoren auf das Auftreten entzündlicher Lungenveränderungen bei Schweinen. *Mh. Vet. Med.* 41, 397–400.

Laanen, M., Persoons, D., Ribbens, S., de Jong, E., Callens, B., Maes, D. and Dewulf, J., 2013. Relationship between biosecurity and production/antimicrobial treatment characteristics in pig herds. *Vet. J.* 198, 508–12.

Loera-Muro, V., Jacques, M., Tremblay, Y., Avelar-González, F., Loera Muro, A., Ramírez-López, E., Medina-Figueroa, A., González-Reynaga, H. and Guerrero-Barrera, A., 2013. Detection of *Actinobacillus pleuropneumoniae* in drinking water from pig farms. *Microbiology* 159, 536–44.

Maes, D., 2012. Subclinical porcine circovirus infection: What lies beneath? *Vet. J.* 194, 9.

Maes, D., Deluyker, H., Verdonck, M., Castryck, F., Miry, C., Vrijens, B. and de Kruif, A., 1999. Risk indicators for the seroprevalences of *Mycoplasma hyopneumoniae*, porcine influenza viruses and Aujeszky's disease virus in slaughter pigs from fattening pig herds. *J. Vet. Med. B* 46, 341–52.

Maes, D., Deluyker, H., Verdonck, M., Castryck, F., Miry, C., Vrijens, B., Ducatelle, R. and de Kruif, A., 2001. Non-infectious herd factors associated with macroscopic and microscopic lung lesions in slaughter pigs from farrow-to-finish pig herds. *Vet. Rec.* 148, 41–6.

Maes, D., Segales, J., Meyns, T., Sibila, M., Pieters, M. and Haesebrouck, F., 2008. Control of *Mycoplasma hyopneumoniae* infections in pigs. *Vet. Microbiol.* 126, 297–309.

Maes, D., Van Soom, A., Appeltant, R., Arsenakis, I. and Nauwynck, H., 2016. Porcine semen as a vector for transmission of viral pathogens. *Theriogenology* 85, 27–38.

Maes, D., Verbeke, W., Vicca, J., Verdonck, M. and de Kruif, A., 2003. Benefit to cost of vaccination against *Mycoplasma hyopneumoniae* in pig herds under Belgian market conditions from 1996 to 2000. *Livest. Sci.* 83, 85–93.

Maes, D., Deluyker, H., Verdonck, M., Castryck, F., Miry, C., Vrijens, B. and de Kruif, A., 2000. Herd factors associated with the seroprevalences of four major respiratory pathogens in slaughter pigs from farrow-to-finish pig herds. *Vet. Res.* 31, 313–27.

Maselyne, J., Adriaens, I., Huybrechts, T., De Ketelaere, B., Millet, S., Vangeyte, J., Van Nuffel, A. and Saeys, W., 2015. Measuring the drinking behaviour of individual pigs housed in group using radio frequency identification (RFID). *Animal*, doi:10.1017/S1751731115000774.

Meyns, T., Vansteelant, J., Rolly, E., Dewulf, J., Haesebrouck, F. and Maes, D., 2011. A cross-sectional study of risk factors associated with pulmonary lesions in pigs at slaughter. *Vet. J.* 187, 388–92.

Mikkelsen, L., Naughton, P., Hedemann, M. and Jensen, B., 2004. Effects of physical properties of feed on microbial ecology and survival of *Salmonella enterica* serovar Typhimurium in the pig gastrointestinal tract. *Appl. Environ. Microbiol.* 70, 3485–92.

Nathues, H., Spergser, J., Rosengarten, R., Kreienbrock, L. and Grosse Beilage, E., 2016. Value of the clinical examination in diagnosing enzootic pneumonia in fattening pigs. *Vet. J.* 193, 443–7.

Neumann, E., 2012. Disease transmission and biosecurity. In *Diseases of Swine*. Zimmermann, J., Karriker, L., Ramirez, A., Schwartz, K. and Stevenson, G. (eds), Wiley-Blackwell, Chichester, West Sussex, UK, 552–624.

Niemi, J., Jones, P., Tranter, R. and Heinola, K. 2016. Cost of production diseases to pig farms. In Proc. 24th IPVS congress, Dublin, Ireland, 7–10 June 2016, 302.

Noyes, E., Feeney, D. and Pijoan, C., 1990. Comparison of the effect of pneumonia detected during lifetime with pneumonia detected at slaughter on growth in swine. *J. Am. Vet. Med. Assoc.* 197, 1025–9.

Opriessnig, T., Meng, X-J. and Halbur, P., 2007. Porcine circovirus type 2-associated disease: update on current terminology, clinical manifestations, pathogenesis, and intervention strategies. *J. Vet. Diagn. Invest.* 19, 591–615.

Opriessnig, T. and Wood, R., 2012. Erysipelas. In *Diseases of Swine*. Zimmermann, J., Karriker, L., Ramirez, A., Schwartz, K. and Stevenson, G. (eds), Wiley-Blackwell, Chichester, West Sussex, UK, 2746–79.

Österberg, J., Vågsholm, I., Boqvist, S. and Lewerin, S., 2006. Feed-borne outbreak of Salmonella Cubana in Swedish pig farms: risk factors and factors affecting the restriction period in infected farms. *Acta Vet. Scand.* 47, 13–22.

Otake, S., Dee, S., Corzo, C., Oliveira, S. and Deen, J., 2010. Long-distance airborne transport of infectious PRRSV and *Mycoplasma hyopneumoniae* from a swine population infected with multiple viral variants. *Vet. Microbiol.* 145, 198–208.

Otake, S., Dee, S., Rossow, K., Joo, H., Deen, J., Molitor, T. and Pijoan, C., 2002. Transmission of porcine reproductive and respiratory syndrome virus by needles. *Vet. Rec.* 150, 114–15.

Patterson, A., Ramamoorthy, S., Madson, D., Meng, X., Halbur, P. and Opriessnig, T., 2011. Shedding and infection dynamics of porcine circovirus type 2 (PCV2) after experimental infection. *Vet. Microbiol.* 149, 91–8.

Pearson, H., Toribio, J. A., Lapidge, S. and Hernández-Jover, M., 2016. Evaluating the risk of pathogen transmission from wild animals to domestic pigs in Australia. *Prev. Vet. Med.* 123, 39–51.

Pedersen, K., Stege, H. and Nielsen, J. P., 2011. Evaluation of a microwave method for dry matter determination in faecal samples from weaned pigs with or without clinical diarrhoea. *Prev. Vet. Med.* 100, 163–70.

Pineiro, C., Morales, J., Doncecchi, P., Dereu, A., Macarrilla, J., Bahnolzer, E., Wuyts, N. and Azlor-Marsinach, O., 2012. In Proceedings of 4th European Symposium of Porcine Health Management, Bruges, Belgium, 102.

Pointon, A., Heap, P. and McCloud, P., 1985. Enzootic pneumonia of pigs in South Australia – factors relating to incidence of disease. *Aust. Vet. J.* 62, 98–100.

Poljak, Z., Dewey, C., Friendship, R., Martin, S. and Christensen, J., 2008. Multilevel analysis of risk factors for *Salmonella* shedding in Ontario finishing pigs. *Epidemiol. Infect.* 136, 1388–400. https://doi.org/10.1017/s0950268807009855

Postma, M., Backhans, A., Collineau, L., Loesken, S., Sjölund, M., Belloc, C., Emanuelson, U., Grosse Beilage, E., Stärk, K. and Dewulf, J., 2016. The biosecurity status and its associations with production and management characteristics in farrow-to-finish pig herds. *Animal* 10, 478–89.

Regula, G., Lichtensteiger, C., Mateus-Pinilla, N., Scherba, G., Miller, G. and Weigel, R., 2000. Comparison of serologic testing and slaughter evaluation for assessing the effects of subclinical infection on growth in pigs. *J. Am. Vet. Med. Assoc.* 217, 888–95.

Ribbens, S., Dewulf, J., Koenen, F., Maes, D. and de Kruif, A., 2007. Evidence of indirect transmission of classical swine fever virus through contacts with people. *Vet. Rec.* 160, 687–90.

Rohrbach, B., Hall, R. and Hitchcock, J., 1993. Effect of subclinical infection with *Actinobacillus pleuropneumoniae* in commingled feeder pigs. *J. Am. Vet. Med. Assoc.* 202, 1095–8.

Saif, L., Pensaert, M., Sestak, K., Yeo, S. G. and Jung, K., 2012. Coronaviruses. In *Diseases of Swine*. Zimmermann, J., Karriker, L., Ramirez, A., Schwartz, K. and Stevenson, G. (eds), Wiley-Blackwell, Chichester, West Sussex, UK, 1821–1914.

Scott, A., McCluskey, B., Brown-Reid, M., Grear, D., Pitcher, P., Ramos, G., Spencer, D. and Singrey A., 2016. Porcine epidemic diarrhea virus introduction into the United States: root cause investigation. *Prev. Vet. Med.* 123, 192–201.

Scudamore, J., Trevelyan, G., Tas, M., Varley, E. and Hickman, G. 2002. Carcass disposal: lessons from Great Britain following the foot and mouth disease outbreaks of 2001. *Rev. Sci. Tech. OIE* 21, 775–87.

Sibila, M., Pieters, M., Molitor, T., Maes, D., Haesebrouck, F. and Segales, J., 2009. Current perspectives on the epidemiology of *Mycoplasma hyopneumoniae* infection. *Vet. J.* 181, 221–31.

Sjölund, M., Postma, M., Collineau, L., Lösken, S., Backhans, A., Belloc, C., Emanuelson, U., Große Beilage, E., Stärk, K. and Dewulf, J., 2016, Quantitative and qualitative antimicrobial usage patterns in farrow-to-finish pig herds in Belgium, France, Germany and Sweden. *Prev. Vet. Med.*, submitted.

Smart, N., Miniats, O., Rosendal, S. and Friendship, M., 1989. Glasser's disease and prevalence of subclinical infection with *Haemophilus parasuis* in swine in southern Ontario. Can. *Vet. J.* 30, 339–43.

Soerensen, D. and Pedersen, L., 2015. Infrared skin temperature measurements for monitoring health in pigs: a review. *Acta Vet. Scand.* 57, 5. DOI 10.1186/s13028-015-0094-2.

Straw, B., 1991. Performance measured in pigs with pneumonia and housed in different environments. *J. Am. Vet. Med. Assoc.* 198, 627–30.

Szabo, J. and Minamyer, S., 2014. Decontamination of biological agents from drinking water infrastructure: a literature review and summary. *Environ. Int.* 72, 124–8.

Thrusfield, M., 2005. The economics of animal disease (Chapter 50), In *Veterinary Epidemiology*, third edition, Blackwell Science, Oxford, UK, 357–67.

Tobias, T., Bouma, A., van den Broek, J., van Nes, A., Daemen, A., Wagenaar, J., Stegeman, J. and Klinkenberg, D., 2014. Transmission of *Actinobacillus pleuropneumoniae* among weaned piglets on endemically infected farms. *Prev. Vet. Med.* 117, 207–14.

Toma, B., Vaillancourt, J., Dufour, B., Eloit, M., Moutou, F., Marsh, W., Bénet, J. J., Sanaa, M. and Michel, P., 1999. *Dictionary of Veterinary Epidemiology*, Iowa State University, Ames, 199–200.

Whitworth, K., Rowland, R., Ewen, C., Trible, B., Kerrigan, M., Cino-Ozuna, A., Samuel, M., Lightner, J., McLaren, D., Mileham, A., Wells, K. and Prather, R., 2015. Gene-edited pigs are protected from porcine reproductive and respiratory syndrome virus. *Nat. Biotechnol.* 2015; DOI:10.1038/nbt.3434.

Chapter 3

Improving biosecurity in poultry flocks

Jean-Pierre Vaillancourt and Manon Racicot, Université de Montréal, Canada; and Mattias Delpont, École Nationale Vétérinaire de Toulouse, France

1 Introduction

Biosecurity is any actions or health plans designed to protect a population against infectious and transmissible agents (Toma et al., 1999). They are based on principles that have been known for centuries. Sanitation measures are detailed in *Deuteronomy*, written about 2600 years ago (Ojewole, 2011); most biosecurity measures found in modern poultry farms appear like analogies of measures prescribed to lepers during the mass of separation in the Middle Ages (Brody, 1974). Yet, we are still struggling today to get people to perform simple tasks that we know work for preventing contagious diseases on poultry

http://dx.doi.org/10.19103/AS.2022.0104.05

farms. For example, in 2011, Racicot et al. identified 44 different errors when getting in and out of poultry barns. Most errors had to do with area delimitation (clean versus contaminated areas of the entrance or anteroom of a barn), boots, and handwashing.

Biosecurity on poultry farms has been categorized in different ways (internal, external; operational, structural; bioexclusion, biomanagement, biocontainment, etc.). Ultimately, all on-farm biosecurity measures have to do with two principles: reducing sources of contamination and separating these sources from healthy flocks. In modern poultry production, a third principle linked to the latter is emerging: the need for a regional approach to biosecurity based on communication within the poultry industry and the organization of movement of people, birds, material, and equipment within any given region. Indeed, the intensification of poultry production has created an environment that may promote the spread of contagious diseases. Any poultry activity involves inherent infectious disease transmission risks, and the amplitude of these risks increases with regional poultry density (Fernandez et al., 1994).

However, no matter how we define or partition biosecurity measures, the recurring challenge is the consistent implementation of these measures by all farm personnel and visitors. Achieving high compliance with biosecurity measures is, therefore, a constant issue that needs addressing in terms of research priority.

2 Reducing sources of contamination: cleaning and disinfection of poultry barns

The decontamination of poultry barns is a multistep process. The first step is the dry removal of organic material. It consists of removing with physical force as much visible organic material as possible: litter, feed, dust, etc. The design of the building and its equipment can significantly affect cleaning (Cerf et al., 2010). For example, the floor of an anteroom with ground equipment will be more difficult to clean on a regular basis compared to one where most equipment is suspended.

The second step is the application of a detergent followed by stripping and drying: some organic materials or mineral deposits are impossible or very difficult to remove using only physical force (Step 1). Step 2 is, therefore, the step in which this material is removed using a chemical process, such as a detergent. It makes it possible to reduce the bacterial concentration of a contaminated surface by a factor of 10-1000 compared to a wash with only clean water (Course et al., 2021). Using a detergent also helps reduce the total amount of water used. Several chemicals influence the effectiveness of detergents: surfactants, wetting agents, dispersants, pH adjusters, and sequestering agents. However, two main categories are recognized: acid and alkaline detergents.

A rule of thumb is that acid detergents remove mineral deposits, while alkaline detergents remove organic waste not removed by dry cleaning (i.e. dried, greasy, or sticky material). The choice of detergent, therefore, depends on the type of material present at the start and the water used. If the cleaning water used is hard (>120 ppm), more frequent use of an acid detergent is necessary (Watkins and Venne, 2015).

When applying the detergent, there are several rules to follow. Firstly, the contact time: the use of detergent in foamy form with the help of specialized equipment is preferred because this makes it possible to optimize the contact time, which should be 10-20 min (Springthorpe, 2000). Second, the temperature of the cleaning water. If hot water is available, it should be used. Water that is too cold slows down chemical reactions. Water at 40°C is best (Böhm, 1998). Third, the flow and pressure. The flow rate and quantity of detergent are greater when applying the detergent, while the pressure is greater during stripping, which is the action following the application of the detergent. When applying the detergent, it is recommended to use a flow rate of 10-20 L/min at low pressure (300-500 psi) and to use approximately 250-500 mL of solution per square meter (Blondel et al., 2018).

If the application of the detergent takes more than 20 min, then the stripping should be done alternately with the application of the detergent. Stripping is the action of removing the rest of the organic debris that the detergent has made accessible. Using clean high-pressure water (1000-3000 psi) at a 45° angle works well for this step, as it adds mechanical force (Blondel et al., 2018). However, it is important to avoid damaging the surfaces; such pressure can damage wooden structures, creating crevices making future decontamination more difficult (Blondel et al., 2018).

Finally, it is important to let the areas dry before proceeding to the next step. Omitting this step may reduce the effectiveness of disinfection in two ways: by dilution and by inactivation on contact with organic material. The effectiveness of a disinfectant is related to its concentration when applied. The presence of water at the surface of what needs to be disinfected dilutes the disinfectant and reduces its effectiveness (Böhm, 1998).

The third step is disinfection, followed by drying. This step should only take place when all organic material has been removed and the areas are dry (Langsrud et al., 2003; Payne et al., 2005). It is important to visually monitor the area that has been cleaned before disinfection (Fig. 1). A nonpermanent marker such as a bright color aerosol hair spray or a tape can be used to identify where cleaning is not satisfactory (presence of organic material); this way, it is not necessary to completely redo the cleaning process for the entire barn. This approach may show over time areas that require special attention.

Disinfection can be carried out by vaporization (liquid or foam) and/or by fumigation/misting. Regardless of the method and product used, it is important

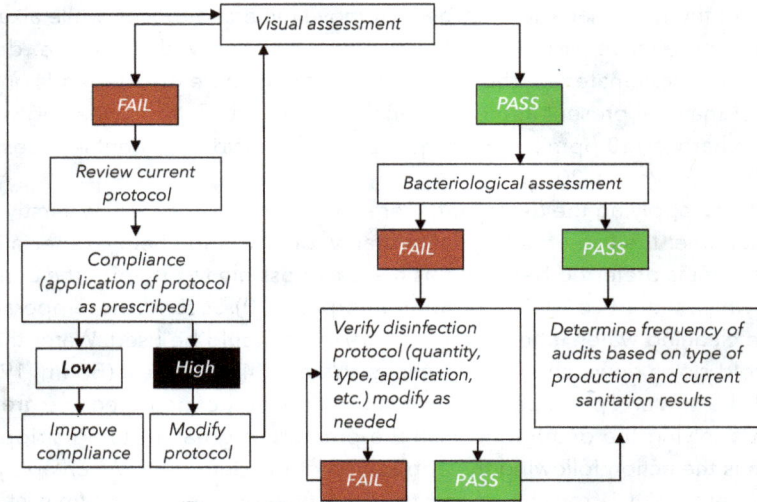

Figure 1 Pathway to assess an on-farm cleaning protocol.

to follow the manufacturer's recommendations for concentration, contact time, product mixture, etc. (Payne et al., 2005). On average, 150–300 mL/m² is applied (Blondel et al., 2018).

The choice of disinfectant depends on several factors: the surface to be disinfected, the pathogens targeted, the hardness and the pH of the washing water, the cost, etc. Generally, the commercial disinfectants available for livestock buildings consist of active ingredients from the following main families: halogenated derivatives, aldehydes, quaternary ammoniums, phenols, and peroxide. Drying the building after the application of the disinfectant is essential, as it prevents the multiplication of the remaining bacteria (Davies and Wray, 1995), and this desiccation period may contribute to reducing further the number of living organisms (Böhm, 1998). In some countries, heating the house (40°C) for 3-4 days has been useful (e.g. preventing the spread of *Mycoplasma gallisepticum* in North Carolina, USA). Finally, a microbiological assessment of the disinfection process may be performed as part of the monitoring (Fig. 1).

3 Reducing sources of contamination: equipment and vehicles

Vehicles and equipment may act as a mechanical vector. Dee et al. (2002) demonstrated that snow contaminated with the Porcine Reproductive and Respiratory Syndrome (PRRS) virus and applied to the undercarriage of a car can contaminate the environment for several kilometers. The inadequate management of vehicles, including where they park on a production site (too close to the barn entrance), has been associated with a greater risk of disease transmission between farms (Guinat et al., 2020). Rose et al. (2000) demonstrated an association between the surface covered by vehicles on a broiler chicken production site and the probability of isolating *Salmonella*. Snow et al. (2010) also identified a link between on-farm transportation and *Salmonella* in egg layer farms. Vehicle-sharing practices between poultry production sites are also associated with the spread of infectious laryngotracheitis and avian influenza (Nishiguchi et al., 2007; Volkova et al., 2012).

Wheel baths are assumed to be useful (Pinto and Urcelay, 2003; Zhang et al., 2013), but there is little evidence that they work. In fact, several arguments have been presented to support their lack of effectiveness. Briefly, wheels rotating at high speed at a high temperature over dry surfaces are not likely to harbor pathogens; even if they have mud on them, passing quickly in a wheel bath does not allow the disinfectant, assuming the solution is fresh, to reach the pathogens; finally, the contact time is too limited for disinfectants to reduce wheel contamination. Pathogens in vehicles are more likely to be present in the loading area and in the cab of the vehicle (Chaber and Saegerman, 2017).

The recommendations made for decontaminating poultry barns essentially apply to the decontamination of vehicles and equipment. When possible, they should be exposed to the sun, which helps the drying process as well as decontamination. Indeed, the ultraviolet radiation in sunlight works as a natural disinfectant (Kuney and Jeffrey, 2002).

4 Reducing sources of contamination: water and feed hygiene

4.1 Water hygiene

Water quality and sanitation would require its own chapter. Briefly, when using chlorine, it is best to assess the oxidation reduction potential (ORP) of water. It measures in millivolts the oxidizing potential of free chlorine residual. A strong oxidizer will effectively destroy microbes. An ORP value of 650 millivolts or greater is required. The goal should be to maintain 2–5 ppm of free chlorine in the water supply. Free chlorine is the residual available for sanitation (Watkins and Venne, 2015). For removing scale and biofilm, it is best to drop the water pH below 5 but not below 4 to prevent equipment damage. A bleach solution might be effective in removing biofilm, but it may also damage equipment. Several disinfectants are available to clean water lines, but some of the most effective and safe products for drinker systems are concentrated, stabilized hydrogen peroxides. Biofilms or any growth of bacteria, molds, and fungus in drinker systems can only be removed with cleaners that contain sanitizers (Watkins and Venne, 2015).

4.2 Feed hygiene and delivery

Basic feed mill biosecurity focuses on tight and closed silos to prevent rodent access, a continuous integrated pest management program, and sanitation procedures, mainly for silos that may have been contaminated with pathogens of importance to poultry. The main one, in terms of public health, is *Salmonella* (Jones, 2011). Li et al. (2012) reported that 8.8% of ingredients of animal origin obtained from three feed mills were contaminated with *Salmonella*, but dust samples had a higher contamination rate of 18.5%. Poultry offal meal and feather meal should be considered high-risk ingredients (Butcher and Miles, 1995). Several chemical options have been tested for the treatment of feeds to control *Salmonella*. Formaldehyde is one of the more frequently used chemicals under commercial conditions (Ricke et al., 2019). Cochrane (2016) provide detailed information on feed mill biosecurity plans.

Basic biosecurity for feed delivery starts with a check with dispatch for routing instructions to avoid passing by potentially diseased production sites or going from one to a disease-free operation. The driver of a clean feed truck must wear clean clothes and have plastic boots, hand cleaning solution, and fly spray available when entering a farm; the driver must never enter barns past the contaminated zone, in other words, must not get on the clean side of the anteroom (Jones, 2011). After all runs are completed, the driver goes directly back to the feed truck area. Trucks must be washed completely and disinfected weekly, unless the last site visited has birds infected with a disease of significance to the industry. In this case, truck decontamination must be performed after this last delivery. Breeder farms are normally visited first before meat bird operations. Delivery to diseased or quarantined farms should be done as the last load of the day (Anon., 2018).

5 Reducing sources of contamination: insect, mite and rodent pests, wild birds and pets

An often-overlooked vector, but with a significant risk of disease transmission, is the common house fly, *Musca domestica*. In 1975, more than 100 different pathogens that could be carried by flies and more than 65 diseases transmitted to humans and animals, such as salmonellosis, had already been identified. They can travel more than 30 km, but the majority of flies are confined to an area of 3 km². They can move quickly over long distances: 1–1.5 km in 24 h, 3–5 km in 48 h, and 6 km in 72 h (Schoof, 1959). Disease transmission can therefore occur between neighboring farms. Flies can also pose a public health problem. In 2009, researchers isolated enterococci and staphylococci in the digestive tract and on the exterior surfaces of flies caught near poultry farms. These bacteria were resistant to antibiotics of importance in human medicine (Graham et al.,

2009). A study showed that 8.2% of the flies tested within 50 m of a farm were positive for *Campylobacter jejuni* (Hald et al., 2004). A study in Japan on highly pathogenic avian influenza H5N1 showed that within a kilometer of the farm, 30% of the flies were contaminated with the same virus as the poultry; at 2 km, the percentage decreased to 10% (Sawabe et al., 2006).

Besides flies, mealworms or darkling beetles (*Alphitobius diaperinus*) are an important concern in disease control. They can transport and transmit microorganisms such as *Campylobacter*, *Salmonella*, infectious bursal disease virus, reovirus, and Eimeria. The mealworm is one of the most abundant insects in poultry farm litter, reaching levels as high as 1000 mealworms per square meter. The risk of disease dispersal can therefore be considerable when spreading litter. In addition, chickens, wild animals, and rodents feed on these insects and can thus infect themselves (Bates et al., 2004; Crippen and Sheffield, 2006). Mainly considering that darkling beetles can maintain *Salmonella* internally during pupation, the next generation of beetles can recontaminate a poultry barn and the following flocks (Roche et al., 2009). Both fly and darkling beetle populations can explode within a few weeks, even with new litter (Fig. 2). Mites (*Dermanyssus gallinae*) have also been associated with recurring site contaminations with *Salmonella Enteritidis* (Moro et al., 2009) and *Salmonella Gallinarum* (Lee et al., 2020)

Insect control requires the application of biosecurity measures for the environment, equipment, visitors, and management of organic materials. It is called integrated pest management. First, the site should always be kept free of unnecessary material since these items can harbor vermin. From the outside, a stone access to the door of the building should be favored. Indoors, rodent traps, or bait should be set in the anteroom and regularly monitored.

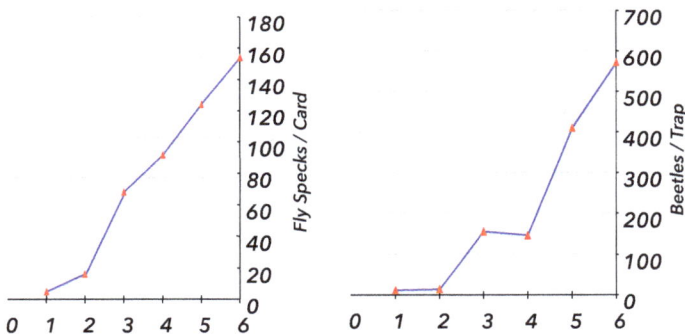

Figure 2 Growth of a fly and darkling beetle infestation in a turkey flock between placement and 6 weeks of age. The fly population was estimated using fly speck card (white card left 24 hours inside the barn and fly defecations counted); Beetle traps where located against the walls inside the barn; the number of beetles where counted after 24 hours.

Insecticides should be applied to the ground, and equipment that must be in the anteroom should be off the floor as much as possible (Villa and Velasco, 1994). It is strongly suggested to rotate between insecticides to increase efficacy and coverage and avoid the development of resistance. Since the treatment is not selective, some recommend the application of insecticides only in areas with high larval density to avoid the destruction of predatory insects, which may be beneficial (Ebeling, 1975).

In addition, feeding and watering equipment should be regularly inspected and repaired to decrease insect populations. The feed delivery system and hoses should be periodically cleaned to prevent insect nesting. To reduce the risk of introducing certain insects, it is also necessary to control employee and visitor traffic and disinfect personal effects and any equipment entering a building. These measures are particularly important when the ectoparasites can survive outside the host for a few days to several weeks. Managing organic materials, such as bird carcasses and manure, is critical. Carcasses provide a substrate and a humidity level (over 90%) favorable for the development of fly larvae, and the manure allows infestation by beetles such as mealworms and their larvae (Allen and Newell, 2005; Barrington et al., 2006).

Rodents can also be found in large numbers on farms as long as water, food, and shelter are available. Rodents are mechanical and biological vectors of several pathogens. Indeed, the same serotypes of *Pasteurella multocida* are isolated in rats and poultry (Curtis et al., 1980). Like rats, mice are involved in the transmission of *Salmonella* (Henzler and Opitz, 1992; Davies and Wray, 1995). Mice are three times more infected with *Salmonella Enteritidis* than samples taken from the environment of a poultry farm. It seems that environmental contamination decreases in the absence of mice (Henzler and Opitz, 1992). Another study supports this view: there are three times more chances that the bacteria will persist in a building when rodents are present, despite washing and disinfecting the building (odds ratio (OR) 3.1; 95% confidence interval (CI) 1.1–3.2) (Rose et al., 2000).

Control of rats and mice is only effective if the barns are rodent-proof, that is, having no openings beyond 1.3 cm (Roy and Brown, 1954). It is also important not to attract rodents to farms by picking up any feed spills, not leaving debris on the site, and regularly pruning vegetation around buildings. Vegetation provides rodents with their basic needs (water, food), protection from predators, and material for building their nests. Appropriate pest management also includes traps or traps containing rodenticides and monitoring to ensure the effectiveness of the program (Axtell, 1999; Corrigan, 2006).

Other wild animals can be reservoirs and vectors of several diseases. Raccoons, along with opossums and foxes, appear to play a role in avian influenza outbreaks. During the 2002 epidemic in Virginia, a significant association was demonstrated between their presence on a farm site and the presence of the avian influenza virus in commercial chickens (OR 1.9; 95% CI 1.0-3.4) (McQuiston et al., 2005). On the other hand, it is more often migratory birds that are involved in the initial influenza outbreaks of an epidemic. Between 1978 and 2000, turkey farms in Minnesota experienced more than 100 introductions of low-pathogenic avian influenza virus from migrating ducks (Halvorson, 2002). Moreover, free-range poultry farms, more exposed to contact with wild birds, showed a higher risk of low-pathogenic avian influenza infection in the Netherlands between 2013 and 2017 (Bouwstra et al., 2017). Several highly pathogenic avian influenza H5N8 epidemics in poultry- and duck-producing regions in Europe between 2017 and 2021 originated from wild birds. Hence, there is clear evidence that wild birds serve as reservoirs, maintaining the virus in several regions over time (Jeong et al., 2014; Verhagen et al., 2021). The mobility of wild birds is a major concern in terms of disease dispersal. They frequently contaminate the environment, vehicles, equipment, feed storage, etc., with their droppings (Davison et al., 1997). It is, therefore, important to limit the access of wild birds and to avoid contact with domestic birds (Axtell, 1999). In that regard, free-range poultry productions require increased attention to biosecurity measures related to the outdoor range (e.g. type of vegetal cover, reducing the accessibility of feed and water to wild birds) (Bestman et al., 2018; Delpont et al., 2020).

It is best to avoid having pets on a farm. Virulent forms of *Pasteurella multocida* have been isolated from the oral cavity of farm cats (Van Sambeek et al., 1995). Strains affecting cats have also been shown to infect turkeys (Curtis and Ollerhead, 1982). There is little evidence of the role of cats in the direct transmission of the bacteria, except in cases of bites (Korbel et al., 1992). However, they can transmit the infection to rats and hunted wild birds, which can subsequently infect domestic poultry. For any type of production, it is therefore inappropriate to use cats for rodent control on a farm (Corrigan, 2006). Dogs, meanwhile, have been identified as a potential carrier of the avian infectious bursal disease virus (Pagès-Manté et al., 2004). The presence of other animals on the site of a broiler farm, including pigs, cattle, sheep, and poultry other than broilers, is strongly associated with a high risk of infection with *Campylobacter* (OR 6.33; 95% CI 1.54-26.00). Cattle are also an important source of introduction of this bacterium in broiler chickens via the farmers' boots (Van De Giessen et al., 1998).

6 Reducing sources of contamination: manure, litter and dead birds

6.1 Manure and litter management

Manure and used litter, and occasionally new litter, can be very contagious material, particularly of enteric pathogens, although it is well documented that a respiratory virus like the influenza virus can persist and spread easily via this route (Duvauchelle et al., 2013; Kim et al., 2018). Kim et al. (2018) reported a reduced risk of contamination when an outside company was contracted for manure removal. However, in Ontario, in 1995, growers who required an outside company to handle used litter were on average eight times more at risk of having a flock infected with infectious laryngotracheitis in comparison to neighbors located less than 2 km from them and having birds at risk at the same moment (Vaillancourt, 1995). Manure has also been identified as a source of contamination for *Campylobacter* (Arsenault et al., 2007).

Therefore, the handling of manure, storage, decontamination (composting, incineration), and removal conditions are of critical importance.

6.2 Disposal of dead birds

In many countries, rendering is the most prevalent approach to dead bird disposal. However, it has been associated with a greater risk of disease transmission in different species, including poultry. For example, in Virginia, United States, farms using rendering instead of disposing of carcasses on site were seven times more likely to have flocks infected with avian influenza H7N2 in 2002 (McQuiston et al., 2005). The proximity of the rendering container to poultry barns may partly explain the greater risk of contamination when renderers visit the site. Another important factor is the absence of an adequate sanitation protocol for personnel visiting the site where the rendering container is located as they return to the barns.

6.3 Incineration

Incineration refers to the combustion of material to the extent that the resulting end products are heat, gaseous emissions, and residual ash. There are three types of dead animal cremation: (1) stationary cremation, (2) air curtain cremation, and (3) open-air cremation. The overriding consideration affecting the use of on-farm incineration is regulatory. In some regions or countries, regulations state that incineration equipment must contain a secondary combustion chamber to reduce particulates (i.e. 'fly ash') and other emissions, such as polluting hydrocarbons and heavy metals (Chen et al., 2004).

Fixed plant units range from business units designed for on-farm animal cremation to large cremation plants. They are normally powered by diesel, natural gas, or propane. The equipment is relatively easy to operate after a short training. Periodic observation, routine maintenance, and ash cleaning are required. Fuel consumption varies depending on incinerator design and loading rate.

Air curtain incineration involves the use of air forced mechanically through a firebox with refractory panels or a trench in the earth. Each type has distinct characteristics that may increase or limit its potential for use as a carcass disposal method. Air curtain technology was developed primarily as a means of incinerating large amounts of combustible waste resulting from land clearing or natural disasters (Ellis, 2001).

The main feature of air curtain incineration is a high-speed 'curtain' of air produced by a fan above an above-ground combustion chamber or in an earth-burning trench. The air curtain serves to contain smoke and particles in the combustion zone and provides better airflow for warmer temperatures and more complete combustion. It was used for the disposal of dead birds during the avian influenza epidemic in Virginia (United States) in 2002 (Brglez, 2003). This is a fuel-intensive process (mainly wood and diesel fuel), but its use may be justified in order to avoid moving dead animals over long distances (Ellis, 2001).

Outdoor incineration is not recommended for several reasons. The downsides include labor and fuel requirements, reliance on favorable weather conditions, potential for environmental pollution, odors, and negative public perception.

6.4 Composting

Several studies indicate that composting appears to be effective in eliminating infectious pathogens endemic to poultry. In principle, large-scale windrow composting would be effective for the disposal of a large volume of carcasses.

There is little information on the fate of prions or sporulating bacteria such as *Bacillus anthracis* when composting cadavers, which prevents it from being considered an adequate method of managing mortalities by the European Union. Deactivation of prions is difficult: it requires exposure to heat from 980 to 1100°C or to alkaline digestion. Neither of these conditions occurs in the composting process. However, there is some evidence that certain enzymes and competition from organisms may have a beneficial effect in reducing the presence of prions during composting (Bonhotal et al., 2014). In contrast, in North America, unless there is a serious suspicion of prions, which is not the case for poultry, composting is considered an excellent means of eliminating the vast majority of infectious pathogens of interest to animal industries and government authorities. In other words, the advantages of composting, including the elimination of the movement of contaminated organic material, are considered far greater than the disadvantages.

Composting is a natural biological process of decomposing organic material in a predominantly aerobic environment. During the process, bacteria, fungi, and other microorganisms break down organic material into a stable mixture (compost) while consuming oxygen and releasing heat, water, carbon dioxide (CO_2), and other gases (Keener et al., 2000). The use of compost for the routine disposal of on-farm poultry carcasses has increased in prevalence in the United States and Canada over the past 30 years (Tablante and Malone, 2006).

Four variables are considered essential for successful composting: (1) moisture content (40–60%), (2) temperature (45–60°C); (3) oxygen concentration (desirable level 10%), and (4) carbon:nitrogen (C:N) ratio (desirable range 20:1 to 30:1) (Keener et al., 2000).

The process essentially takes place in two phases – a primary thermophilic phase (temperatures up to 70°C) and a secondary mesophilic phase (usually 30–40°C) (Kalbasi et al., 2005).

Temperature and temperature maintenance are important factors in the destruction of infectious pathogens. A temperature of 54°C for 3 days, typical of carcass composting, should kill all pathogens except spores and prions (Sander et al., 2002).

To keep costs down on the farm, composting of carcasses is usually done by producing a static pile or heap that does not involve specialized mixing, crushing, turning, aeration, and screening equipment. The degradation of the carcass is initiated by natural anaerobic bacteria in the cadaver and by aerobic bacteria on the exterior surfaces. Odorous gases and liquids diffuse into drier, more aerobic plant materials, where they are ingested by microorganisms and broken down into simpler organic compounds and ultimately into CO_2 and water (Keener et al., 2000). The success of this operation relies on the careful construction of a layered heap using appropriate amounts of plant-based covering material below, between, and above the carcasses. Characteristics of effective toping materials include water-holding capacity, gas permeability or porosity (oxygen for microbial activity), biodegradability, wet strength, and sufficient carbon. These physical characteristics determine the ability of toping materials to absorb excess liquids, preventing the release of leachate and odors (King et al., 2005).

A variety of plant-based residues have been used as toping materials, including sawdust, wood chips, ground corn stalks, straw or ground hay, oat or peanut hulls, poultry or livestock bedding, dry manure, etc. (Keener et al., 2000; Glanville et al., 2006).

Turning a stack may be necessary to break up wet areas and to introduce more oxygen and moisture, if needed, to reactivate aerobic microbial activity and stimulate a secondary cycle of heat production. Once the secondary heating cycle is complete, soft tissue decomposition is usually complete and the compost is stable enough to be stored prior to field application. Based on a review of the literature, Keener et al. (2000) concluded that decomposition times are largely a function of cadaver mass, and they published weight-based prediction equations for the duration of the primary and secondary composting cycles. In practice, most of the compost is turned only once or twice. Turning accelerates the decomposition of the carcasses, but it is not essential if the carbon source used to cover the carcasses is sufficiently permeable for the diffusion of oxygen into the heap (Glanville et al., 2006).

On-farm composting is usually done in open bins or swath piles. Three-sided bins are typical, with the open end allowing access for placement, turning, and removal of compost using a tractor. Permanent structures are built with treated lumber or concrete and are usually built on a concrete platform to provide a firm work surface. Windrow composting involves the construction of long, narrow piles having a parabolic or trapezoidal cross section. Due to their shape, windrows have a large exposed surface that encourages aeration and drying (Mukhtar et al., 2004). Since the dimensions of the windrows are not constrained by the walls, their dimensions can be adapted to any size and number of carcasses, which makes them particularly useful following the mass slaughter of animals.

Although windrows do not require the construction of a structure to contain the compost, a low permeability base is recommended to avoid contamination by leachate from the underlying soil (e.g. concrete or asphalt; gravel lined with fabric plastic or geotextile; compacted soil) (Keener et al., 2000).

7 Separating healthy birds from sources of contamination: zoning production sites

Basically, zoning is about preventing the contact between susceptible birds and microbes. The barns where the birds are located must be the most restricted zone (restricted access zone or RAZ); the farm where we find the barns is a controlled access zone (CAZ) (Fig. 3). Everything else around these two zones must be considered an environment that is source of contamination (Anon., 2018).

7.1 Preserving the integrity of the controlled access zone

All farm personnel must avoid live bird markets and be in contact with any other poultry. The CAZ should be clearly marked or understood by those who must access the site. A fence is best when economically doable. The main point is that only essential vehicles, personnel, and equipment should get within the CAZ.

An integrated pest management program, including cleaning out vegetation for at least 5 m around the barns, is paramount to removing potential carriers of disease (Vaillancourt and Martinez, 2001). Standing water must be avoided on the premises. It is a breeding environment for insects, and it attracts wild birds that may be carriers of infectious pathogens such as avian influenza. Feed spills

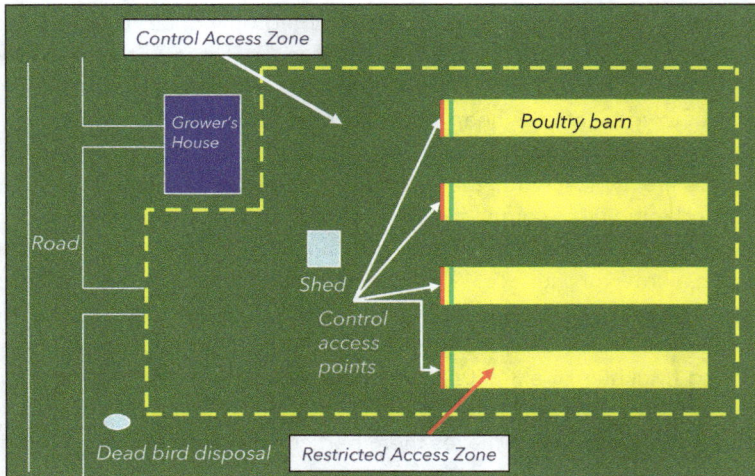

Figure 3 Control access zone (area where poultry traffic and equipment are controlled; delimitation can be virtual or physical (e.g. fence); and restricted access zone (area where birds are located). The control access points are the doors, and ideally the anteroom, providing access to the flock.

should also be avoided for the same reason. If rendering is used to dispose of dead birds, the disposal area cannot be part of the CAZ (Anon., 2018).

Finally, equipment and vehicles must be washed and disinfected, most importantly, all equipment used on more than one farm. In this case, it is best to clean and disinfect when leaving the farm and prior to use on the next farm (Anon., 2018).

7.2 Protecting the restricted access zone

The purpose is to limit access to essential visitors. When a visit is needed, it is important to make sure that visitors have taken a shower and changed clothing and footwear before visiting. Even if they have not been on another farm before, they may still have had an opportunity to get contaminated on their way to the production site. Employees and even the farm owner should consider themselves visitors when they leave the facilities and return. It is best for them to wear farm-specific clothing (laundered each day or supplied at the farm) and boots (Anon., 2018).

7.3 Anteroom: managing barn entrances

The separation between the RAZ and outside must be clearly delineated. When a shower is present, the shower itself is the line separating the two. In the absence of a shower, it is important to delineate the zones with at least a line. Anterooms separated into only two zones (external or contaminated and internal or clean) can be problematic because cross-contamination between the two zones is frequent (Figs 4 and 5). Therefore, keeping the barn entrance free of any organic matter will prevent cross-contamination of these two areas. However, it is difficult to achieve without a physical barrier between the zones. In addition, when designing an anteroom, it is recommended to have separate drainage for each area. Finally, an easier and more effective anteroom design is one with three zones (Fig. 6). Often known as the Danish entrance, the three-zone design offers an area between the contaminated and clean zones where visitors and personnel can wash their hands. This step, while transitioning between the contaminated and clean zones, is important because it ensures that hands are decontaminated before handling boots, clothing, or equipment on the clean side of the anteroom (Anon., 2018).

In an anteroom, porous surfaces (concrete, wood) are generally more prevalent than smooth surfaces (plastic, metal). For detergent application on porous surfaces, it is recommended to use a low-foaming detergent. For disinfection, the use of fumigation (0.5 μm particle) might be preferable for its ability to reach corners and penetrate cracks (Battersby et al., 2017), although research is inconclusive on this point.

The type of soil also has an impact on the sanitation of an anteroom. For example, a cracked concrete floor, a porous soil where water stagnates easily, will be much slower and more difficult to disinfect compared to a floor covered with flat epoxy. Stagnant water is a source of multiplication for microorganisms and should be given special attention.

Figure 4 Equally dirty contaminated (outside) and clean (inside) zones separated by a red line in an anteroom.

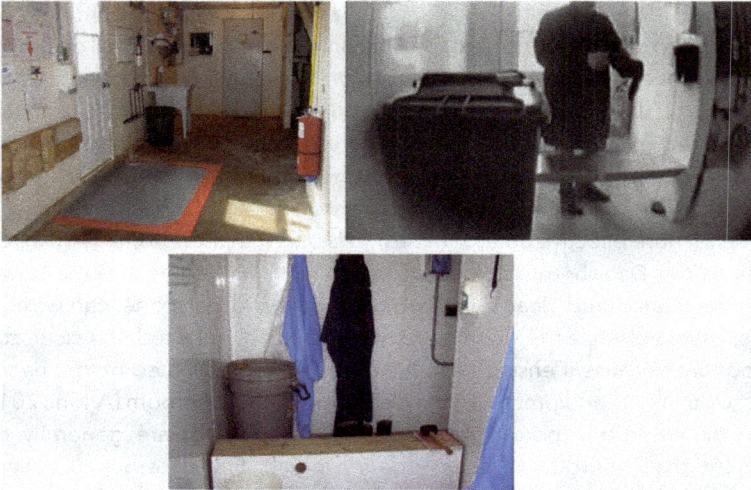

Figure 5 Examples of two-zone barn entrances. Note that a hydroalcoholic solution dispenser for hand decontamination is located at the level of the separation between the zones. The red line design is less effective for compliance than the ones with a bench.

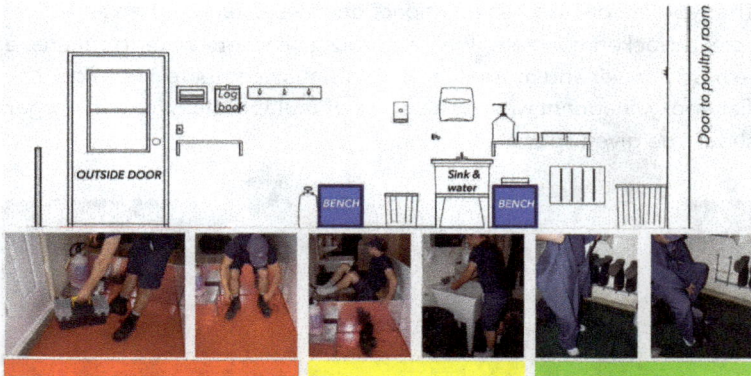

Figure 6 Layout of a three-zone barn entrance and the steps that must be followed to pass from the red zone (contaminated area) to the green zone (clean area) via the yellow zone (intermediary area for hand washing).

7.4 Footbaths

The effectiveness of a footbath is influenced by several factors. The two major factors are the presence of organic matter (on the boots or in the footbath; Fig. 7) (Amass, 1999; Amass et al., 2000; Rabie et al., 2015) and the often insufficient contact time. Other factors also influence the effectiveness of footbaths: the type and concentration of the disinfectant (Amass et al., 2000; Hauck et al.,

Figure 7 Examples of footbaths left outside and rendered ineffective due to a large amount of organic material.

2017), the location (Fig. 8), the pH and the hardness of the water, and the material of the container used (e.g. metal container). For example, if a footbath is left outside, the disinfectant may be inactivated by UV rays from the sun (e.g. iodine), diluted by rain, and affected by warm (evaporation) and cold (freezing) temperatures (Rabie et al., 2015).

Despite all these issues, some epidemiological studies have shown a protective effect of footbaths, especially in connection with contamination with *Campylobacter* spp. and with avian influenza H9N2 (Chaudhry et al., 2017). However, these results are not supported by other studies (Refrégier-Petton et al., 2001). Allen and Newell (2005) report a lower efficiency of footbaths compared to using boots dedicated to each barn.

Beyond epidemiological studies, other studies have evaluated the direct effectiveness of footbaths against infectious pathogens under different conditions (contact time, presence of organic matter, wet or dry prewash, type of disinfectant, etc.). In studies demonstrating good efficacy, Dee et al. (2004) reported that a sodium hypochlorite footbath (bleach) allowed the complete elimination of the PRRS virus with a contact time of 5 s in the disinfectant solution. It is important to note that the footbath was replaced after each use (single use), which is not done in practice. Amass et al. (2006) reported a slight reduction, insufficient to be considered significant, in bacterial contamination on lightly soiled boots when passed over a disinfectant mat. Morley et al. (2005) reported a 67–78% reduction in bacterial load after 7 min of contact time with the disinfectant solution on lightly soiled boots. Finally, Dunowska et al. (2006) reported a reduction in bacterial load of 95.4–99.8% after 10 min of contact on lightly soiled boots. The studies reporting certain effectiveness can thus

Figure 8 Examples of footbaths positioned in such a way as to prevent users to march forward as they should between two zones.

be criticized because the reduction of the microbial load is inadequate, the solution needs to be replaced after each use, or the contact time required is unrealistic in practice.

Several studies have confirmed the inadequacy of traditional footbaths as a biosecurity measure. Hauck et al. (2017) reported that quaternary ammonium and quaternary ammonium-glutaraldehyde footbaths were not able to eliminate highly pathogenic (H5N8) and low-pathogenic (H6N2) influenza. Several studies have demonstrated that the presence of organic matter prevents the effective disinfection of boots, regardless of the contact time (Amass et al., 2000; Rabie et al., 2015). In fact, the mechanical action of a brush to remove organic matter from the boots is even more effective than walking through a footbath without prewashing. It is important to note that a footbath can even be a risk factor (Amass et al., 2000) because it provides a humid environment favoring bacterial growth and sharing of pathogens between treated boots. Having this in mind, Owen and Lawlor (2012) compared phenol- and quaternary ammonium-filled footbaths to dry footbaths filled with a dry bleach powder. The average residual life of the liquid footbaths was less than 2 h and did not significantly impact the bacterial count on the boots. By contrast, the dry bleach footbath reduced the bacterial count by around 98% and had a residual life of 14 days.

7.5 Downtime

Downtime is the period when a poultry barn or an entire production site is without a flock. The concept is to remove birds in order to apply sanitation measures. Repopulation follows a few days to a few weeks later. This is called all-in/all-out (Anon., 2018). This approach has a protective effect on *Salmonella* contamination (Snow et al., 2010) and avian influenza (Kim et al., 2018). In a study on coliform cellulitis in chickens in Canada, the longer the downtime, the lower was the prevalence of cellulitis-related condemnations at slaughter (Elfadil et al., 1996). Chin et al. (2009) reported that an extended downtime of 30–91 days contributed to the regional control and eradication of infectious laryngotracheitis in California. Field experts participating in a Delphi study indicated that a 14-day downtime between meat bird flocks was the norm, with 21 days between breeder flocks (Vaillancourt and Martinez, 2001).

8 Separating healthy birds from sources of contamination: hatchery

There is more to hatchery biosecurity than just sanitation. Hatchery location, design, accessibility, workflow, and pest control all complement a stringent sanitation program to ensure good biosecurity. The temperatures and humidity required for hatching are ideal for the growth of bacteria and molds. Problems coming from the breeders can be amplified, and new ones introduced (Kim and Kim, 2010).

Embryos and newly hatched chicks can offer little disease resistance. In other words, the pressure of infection required for affecting their health and viability is much lower than for older birds. It is, therefore, essential to provide them with a clean environment during their incubation and hatching (Bennett, 2017).

The more a hatchery is located in close proximity to other poultry traffic, the more vigilant one should be about issues such as pest infestation, visitors, and ventilation-related pathogens. It is important to recognize that a lot of poultry in the area increases the risk of transmission of infectious diseases. Therefore, it is best if no poultry-related activities (feed mill, farms, live-haul routes, etc.) are conducted next to the hatchery (Bennett, 2017).

8.1 Personnel

All employees must park in a designated area. They should not be visiting other poultry facilities. On the rare occasion when this might occur, they would have to take a shower and use clean clothes before reentering the hatchery. They should use protective clothing, including boots, only used at the hatchery. If a shower in-shower out facility is available at the hatchery, it must be used consistently and never by-passed. At the very least, all employees will be expected to have taken a shower at home before coming to work, they must come to work wearing clean clothes, and they will all be required to thoroughly wash their hands when arriving at the hatchery. Therefore, a handwashing station with antibacterial soap should be readily available. Handwashing is essential before, after, and between egg and chick or poultry handling operations (e.g. setting, transfer, candling, sexing, vaccinating, packing, etc.) (Thermote, 2006; Bennett, 2017).

8.2 Hatchery workflow

The work and traffic flow should follow the same route as the hatching egg. Modern hatcheries should be designed with ventilation systems to prevent cross-contamination of the different areas of the building. The benefits of such systems will largely be negated if employees are allowed to move freely back and forth between separate areas of the building. If it is needed to 'break' the unidirectional workflow, it is important for all employees to wear clean hatchery clothing and to change it as necessary. The hatchery workflow is as follows: egg receiving area, egg holding area, egg cooler, setters, hatcher rooms, tray dumping, chick processing area, chick holding area, and chick loading area (Mauldin, 2002).

The key is to minimize contamination from one room to the next. Positive pressure rooms are important in critical areas so that contamination will not be drawn in through an open door. Doors help stop cross-contamination between rooms (Mauldin and MacKinnon, 2009).

8.3 Egg delivery

The driver cannot enter the egg holding room at the hatchery, in particular, if several breeder farms are visited. After delivery, the egg truck is washed and disinfected. Clean and disinfected trays and trolleys are returned from the hatchery to the breeder farms (Bennett, 2017).

8.4 Egg sanitation

Eggs are not laid in a sterile environment. Even before that, eggs are exposed to many microbes. The vent of the hen is a contaminated 'delivery environment'. Dirty nests, or if eggs are laid on the floor, can also contribute to egg contamination. Eggshell surfaces must be kept dry. Without humidity, most microbes will have a hard time multiplying or penetrating the egg's natural defenses. Therefore, it is important to avoid the sweating of eggs, such as by moving them from a warm to a cool environment. Bacteria inside an egg will use some of its nutrients and may affect the growth or even kill the embryo. Contaminated eggs in incubators and hatchers can also serve as a source of contamination for the other eggs and newly hatched birds.

Early embryos can also be affected by chemical vapors. So the choice of sanitizing products in the egg reception and storage areas is important (please see the following text). Eggs that have been disinfected on farm can be moved immediately to setter trays for storage. Eggs of unknown sanitation status normally should not be allowed in the hatchery. If the decision is made to allow them, they must be disinfected before setting. If dirty eggs are found (obvious presence of fecal material, litter, and feathers), they should be removed and held for breeder management to review.

Egg sanitizing (Ernst, 2004): It must be done properly; otherwise, the result may be a greater degree of egg contamination. It is important to keep sanitizing water hotter than the eggs (43-49°C). The sanitizing solution must contain a detergent sanitizer. It is best to use a washer that does not recirculate water.

Chlorine and quats (quaternary ammonium) have been excellent disinfectants. Quats offer residual activity, are safe for the operator, are relatively cheap, and are compatible with antibiotic dipping. Quats are safe for hatching eggs up to 10 000 ppm but are effective as a disinfectant at 250 ppm (large safety margin) with 10 ppm ethylenediaminetetraacetic acid. It is best to keep the solution at a pH of 8 for optimum efficacy. This can be achieved by using sodium carbonate (Rodgers et al., 2001).

Hatching eggs can also be sanitized by fumigation using 1.2 mL of formalin and 60 g of permanganate sulfate per cubic meter in a disinfection cabinet. Other products may be used for fumigation, and new technologies are also considered, such as pulsed ultraviolet light (Cassar et al., 2020) and low-energy electron beam (Steiner, 2020).

8.5 Building and equipment sanitation

Washing and disinfection of equipment are a critical component of hatchery biosecurity. A thorough cleaning of the area (setters, hatchers, floors, chick-go-rounds, vaccinators, etc.) is essential before a disinfectant can be applied. Organic material (fluff, blood, shells, and droppings) reduces the effectiveness of disinfectants. Therefore, washing is paramount.

8.6 Setters

Fog a disinfectant after eggs have been set. Multistage setters should be fogged each time new eggs are set or transferred. When setters are emptied, a thorough cleaning and disinfection are required. Like for any other areas, all debris must be removed as part of the cleaning process.

8.7 Hatchers

The hatcher is the main source of organic contamination in the hatchery (eggshells, unhatched eggs, dead chicks/poults, fluff, droppings). It is, therefore, important to use equipment handling and sanitation procedures to minimize contaminating the other areas of the hatchery.

8.8 Bird processing equipment

This equipment requires a very rigid sanitation schedule to avoid bird quality problems (Mauldin, 1983). Special attention should be paid to the vaccination equipment (syringes, vaccine containers, and hoses washed with a nonresidual disinfectant such as alcohol or chlorine).

Chick-Go-Rounds and/or chick/poultry conveyor belts and other processing equipment can contaminate day-old birds. At greater risk are chicks or poults that may have been pulled early. If their navel is incompletely closed and comes in contact with a dirty or contaminated conveyor belt, there will be many cases of infection. Therefore, it is important to clean and disinfect conveyor belts after the chicks/poults from each breeder flock are processed.

8.9 Ventilation

The hatchery ventilation system plays an important role in preventing contamination (when functional and well designed) or by being the source of the problem (e.g. aspergillosis contamination). Ideally, each room of the hatchery should have its own separate ventilation system so air does not move from one room to another. This way, the ventilation system will not introduce contaminated air from other sections of the hatchery or from outside. The system must provide clean, fresh air to the hatchery and the incubators at all times. Hatcheries should be designed to exhaust air away from intake outlets, so contaminated air is not recycled into the hatchery (Mauldin, 1983).

Maintenance of the evaporative coolers is critical. Filters must be cleaned and disinfected routinely. Disinfectants can be added to the water in the evaporative coolers, especially when starting them off in the spring.

8.10 Monitoring

Records must be kept that the biosecurity program is in place and implemented correctly and continuously. The hatchery should have a quality control program to monitor incoming eggs visually and microbiologically. A continuous monitoring program of the building and equipment must be in place to determine the microbial populations in the hatchery (Racicot et al., 2020). Samples may be collected from the tray wash area, air intakes and outlets, filters, evaporative coolers, setters, hatchers, air in chick holding and egg storage rooms, chick conveyor belts, water source to hatchery, vaccination equipment, including vaccine and diluent (Mauldin, 1983). Samples must be tested for bacteria (coliforms, *Salmonella*) and fungus.

9 Separating healthy birds from sources of contamination: regional biosecurity

The relationship between proximity to a poultry production site and the probability of contamination has been well established. When the distance between two egg layer farms increases by 1 km, the risk of colibacillosis is six times less in Belgium (Vandekerchove et al., 2004). Egg layer farms are also twice less likely to be contaminated with *Salmonella* if the nearest farm is at least 1 km away (Snow et al., 2010). The risk of infectious bursal disease (Gumboro) transmission was higher when the susceptible flock was less than 20 km away from an infected flock. Poultry traffic, wild birds, and airborne transmission might have contributed to such disease transmission in Denmark (Sanchez et al., 2005). In Australia, East et al. (2006) demonstrated that the risk of Newcastle disease infection is reduced threefold in egg layer farms if the nearest farm is at least 10 km away; fourfold for breeder farms if the nearest poultry farm is at least 1 km away, and threefold in broiler chickens if at least 500 m away. A poultry farm located less than 1 km away from an avian influenza-infected flock is about 35 times more at risk of infection compared to a farm located at least 10 km away (Boender et al., 2007). Commercial poultry farms infected with the infectious laryngotracheitis virus were 36 times more likely to be within 1.6 km from a backyard flock than disease-free farms (Johnson et al., 2005).

In Australia, veterinary authorities, in collaboration with the poultry industry, have established guidelines for the geographic distribution of poultry farms (Table 1). Although these are not yet mandated by law, they were determined and accepted collectively based on scientific evidence. Note that these are standards for regular poultry operations.

The vast majority of infectious diseases are horizontally transmitted. Cross-traffic between farms and the joint use of equipment are certainly significant

Table 1 A guide on biosecurity buffer distances[a]

Farm type	Species	Buffer (m)
New farm	Fowls/turkeys/other avian species, e.g. ratites, quail	1000
Units in large farm complex	Fowl/turkeys or other avian species	200–500
Farm complexes	Fowl/turkeys or other avian species	>2000
Breeder farms	Fowl/turkeys or other avian species	2000–5000
Duck or waterfowl farms	Duck, waterfowl	5000

[a]The buffer is measured from either the nearest barn walls for older type building or from the centroid of the mechanical ventilation system of the newer tunnel ventilation barns. Reference: http://new .dpi.vic.gov.au/notes/agg/poultry--and--other-birds/ag1155-biosecurity-guidelines-for-poultry -producers

modes of disease transmission (Guinat et al., 2020). With potential vectors such as flies and darkling beetles, known vectors of avian influenza, disease control efforts in regions of high farm concentration cannot rely solely on farm-level measures. In a competitive environment, it is tempting to limit data sharing. Although liability will always be a concern, pointing fingers has never been an effective disease control strategy, and poultry organizations sharing a region must also share the necessary information needed to prevent and contain significant contagious diseases. A regional approach to biosecurity includes establishing separate geographical working zones for service personnel. When an important disease is suspected on a farm, a self-imposed quarantine matched with communication with key personnel will go a long way to prevent a major epidemic (Vaillancourt et al., 2018).

Communication remains central to the success of regional biosecurity, especially when an outbreak occurs. In 2005, during an outbreak of infectious laryngotracheitis in California, two integrators coordinated to perform an extended downtime in a region and to set up biosecurity audits. Cooperation and communication between companies have helped eradicate the disease. The importance of communication is also highlighted in a 2001 survey of 72 North American veterinarians on the importance given to certain elements of biosecurity (Vaillancourt and Martinez, 2001).

10 Biosecurity compliance

Biosecurity compliance is defined as the behavior of a person coinciding with the recommendations of professionals in the control and prevention of infectious diseases. The measure of compliance is reported as the ratio of the number of recommendations applied to the number of recommendations prescribed; in other words, the ratio of the number of biosecurity measures applied to the number required. In practice, this is expressed as a percentage.

The constant application of biosecurity measures is essential for the success of any type of animal production. However, compliance with biosecurity measures is sporadic and variable regardless of the type of production, including pig and poultry productions (Losinger et al., 1998; Pinto and Urcelay, 2003; Boklund et al., 2004; Ribbens et al., 2008).

Similar observations are made in human medicine. Indeed, although compliance with hygienic measures among physicians is relatively high (80%) during high-risk procedures, this is not the case (30%) for low-risk procedures. In addition, it is interesting to note that compliance with medical recommendations by patients (33-54%; Haynes et al., 1979) also corresponds to what is observed in commercial poultry farms (Racicot et al., 2011).

There is a weak correlation between reported compliance and observed compliance ($r = 0.21$) for handwashing (O'Boyle et al., 2001). In veterinary medicine, few studies report compliance as observed. When it does, the results are worrying. For example, a procedure to register visitors to turkey farms in North Carolina was filmed on three farms shortly after the procedure was put in place. Compliance ranged from 7% to 49% (Vaillancourt and Carver, 1998).

A study in Quebec on compliance with biosecurity measures in the anteroom of poultry production buildings on eight farms was carried out using hidden cameras (although stakeholders on the farm were informed that they would be filmed). A total of 44 different errors were observed from 883 visits made by 102 different people. On average, four errors were recorded per visit. The maximum number of errors made by a person during a visit was 14. Twenty-seven of the 44 errors (61.4%) were related to measures relating to the respect of the zones (clean or internal versus contaminated or external), 6 for boots (13.6%), 5 for handwashing (11.4%), 3 for coveralls (6.8%), and 3 for visitation records (6.8%). The nature and frequency of errors suggest a lack of understanding of biosecurity principles (Racicot et al., 2011).

Since compliance with biosecurity is generally weak, it is essential to establish strategies to improve the implementation of biosecurity measures. A study evaluated the value of audits and visible cameras on compliance with the biosecurity measures required when entering and leaving poultry barns of 24 farms in Quebec. The evaluation was done initially during the first 2 weeks after the intervention, as well as 6 months later for another evaluation lasting 2 weeks. Nearly 2800 visits by 259 different people were recorded on video. The results showed that bimonthly audits had no impact on compliance in the medium term, 6 months after the start of the project. It should be noted, however, that the auditor was not in a position of authority. The visible cameras had an impact on the change of boots (OR = 9.6; 95% CI 1.9-48.4) and respect for areas (contaminated versus clean) during the visit (OR = 14.5; 95% CI: 1.2-175.1) for the short-term period. However, 6 months later, compliance declined and was no longer significantly different from control farms. The duration (> 5 min) and the time of the visit (morning), the presence of the producer or an observer (sometimes negative effect), the design of the anteroom (in relation to the ease of application of the measures), the number of buildings (more than five buildings), the number of biosecurity measures requested, the type of boots worn (plastic boots), and being a member of the grower's family (negative effect) were significantly associated with biosecurity compliance. The same study also showed that certain personality traits were associated with compliance, as well as the number of years of experience in poultry production and the level of education of the participants. For the number of years of experience, the relationship was not linear. It is possible that early-career circumstances (e.g. avian influenza epidemic) could have a long-term impact on compliance (Racicot et al., 2012b).

Delpont et al. (2020) attempted to determine how sets of socio-psychological factors (i.e. knowledge about biosecurity and transmission of avian influenza, attitudes, personality traits, social background) affect the adoption of on-farm biosecurity practices. The study was carried out as part of 127 visits to duck farms in southwestern France. The factorial analysis of mixed data and the analysis of hierarchical grouping identified three groups of producers with different socio-psychological profiles. The first group was characterized by minimal knowledge, negative attitudes toward biosecurity, low social pressure (defined as a concern to preserve the avian influenza free status of the farm in order not to tarnish one's reputation), and a low level of conscientiousness (conscientiousness is a fundamental personality trait that reflects the tendency to be responsible, organized, hardworking, goal-oriented, and adheres to standards and rules). The second group was characterized by greater experience in poultry production, increased stress (excessive nervous tension), and social pressure. The third group was characterized by less experience in poultry production but better knowledge and positive attitudes toward biosecurity, increased self-confidence, and a focus on action. The first group had significantly lower adoption of biosecurity measures than the other two groups.

The measures implemented on farms vary greatly, depending on the animal species, type of production, region, stocking density, and possibly regional health status. For example, a Canadian survey (Young et al., 2010) of 642 broiler breeder owners indicates that 13% of respondents do not allow access to visitors. Of the rest, 36% require them to wash their hands and 50% require additional protective clothing to be worn before entering a barn. The main reasons for not requiring visitors to wash their hands were the lack of perceived need (20%), the time required by this measure (20%), and the lack of facilities (18%). Almost a third of farm owners also did not require the use of coveralls because of the time required to put them on, and 20% found it sufficient to wear only boots.

The known risk of disease transmission could favor the better implementation of measures (Dorea et al., 2010). The implementation of biosecurity measures also depends on the individuals being questioned. Indeed, producers and technical staff responsible for health monitoring do not always agree on what should be done and even what is done on a given farm. For example, in a study carried out in Canada, a weak to slight agreement was observed when these technicians were asked about the restrictions required by the farm owner to have access to the farm and their responses were compared to those of these owners (Nespeca et al., 1997). It is therefore important to provide each employee with written plans for the required biosecurity measures and to ensure a continuous training program (England, 2002).

10.1 Obstacles to compliance

Several reasons are given to explain the lack of compliance with biosecurity measures. Lack of knowledge or understanding of measurements is the main reason (Lotz, 1997; Barcelo and Marco, 1998; Sanderson et al., 2000). Lack of communication, time, incentives (positive and negative) to follow the rules, lack of audit programs, apathy or denial of potential risks, and economic constraints are also considered to be important factors (Vaillancourt and Carver, 1998). Communication is particularly necessary during epidemics. In 2005, during an outbreak of infectious laryngotracheitis in chickens in California, USA, two integrated companies coordinated to carry out an extended downtime and implement extensive audits. Cooperation and communication between companies have helped eradicate the virus (Chin et al., 2009). Gunn et al. (2008) also underline the importance of better collaboration between producers, veterinarians, and technical staff working on farms. To ensure good communication, the messenger is essential, but the content of the message is decisive. Several private and public organizations produce training materials for producers, but their content varies widely. This lack of harmony between training programs, and the resulting confusion, likely contribute to the lack of application of biosecurity (Jardine and Hrudey, 1997; Moore et al., 2008).

For a given disease, the perceived risk of infection influences the implementation of biosecurity measures. Unfortunately, the perception of risk differs greatly between producers, making consistency in the application of biosecurity measures difficult. The perception of risk is strongly linked to the understanding of the principles of biosecurity and the regional incidence of diseases. Perception of risk is also influenced by the intensity of traffic on the farm. Thus, producers of small farms or family farms often consider themselves less at risk compared to producers of intensive commercial livestock farms (Larsen, 2009).

In addition to the perception of risk, other individual factors may influence the application of recommendations. These are the attitudes, personality traits of the individual, experience with the disease, level of education and personal beliefs about animal health, incidence, prevention, and disease control (Buckalew and Sallis, 1986; Strömberg et al., 1999; Delabbio et al., 2005).

10.2 Intervention strategies

Haynes et al. (1979) describe different strategies to improve adherence to medical recommendations. Reminders of recommendations increase compliance from 24% to 70%. Thus, proper instructions with clear information and repeated feedback to stakeholders should improve compliance.

When it comes to handwashing, studies show that increasing accessibility to washing stations, training programs, and frequent feedback significantly increase compliance. This can go from less than 20% to more than 70% after the establishment of a training program and a monitoring system with frequent feedback (Tibballs, 1996; Colombo et al., 2002). However, compliance declines over time. This deterioration appears to be multifactorial and probably associated with the lack of a continuing education program (Conly et al., 1989).

10.3 Motivation of farm personnel

Motivation is another important element. Compliance with industrial safety rules is intimately linked to the motivation of employees to comply (Vroom, 1994). Clear goal statements and performance feedback have an optimal motivational effect when these two elements are combined. To increase motivation toward desired behavior such as compliance with biosecurity measures, the ability to achieve goals and a sense of self-efficacy are important (Bandura and Cervone, 1983). Self-efficacy is defined as 'the ability to implement the behavior essential to obtain a given result' (Pervin and John, 2001). According to Bandura and Cervone (1983), 'Our ability to deal with a situation and control the outcome is the key element that actually influences behavior. It is by manipulating the feeling of self-efficacy (e.g. by means of feedback announcing to the person that, compared to the performance obtained by others, their own performance is very good), rather than by giving feedback regarding a risky behavior, that a change in behavior is likely (Pervin and John, 2001).

Millman et al. (2017) have clearly demonstrated the perverse effect of a contradiction between what is required and what can be achieved. They studied the behavior of poultry catchers who had to follow several biosecurity procedures when the time allowed to do so was not sufficient. This situation led catchers to openly question the protocols and the need to respect biosecurity measures. Their noncompliance with the requirements was then seen as resourcefulness in the face of unrealistic conditions. Timmermans and Berg (1997) also stress the importance of 'local universality'; that is, standardized practices, such as biosecurity measures, can only be universal (achievable in different places and times) if they can be adapted locally.

10.4 Training program

A training program should include an evaluation of the participants before the training (theoretical and practical parts) and at the end. It is necessary that the trainer be credible and have the confidence of the participants so that the message is delivered well. Participants should also leave the session with a copy of the training material. A website accessible to participants will include materials, as well as program updates, upcoming training schedules, and links to other sites providing information on biosecurity. Conly et al. (1989) showed that training programs are not sufficient to maintain long-term compliance. Feedback and reinforcements (incentives) are needed. Reminders in various forms are also useful, such as messages in agricultural journals, mailings, contests and awards at annual meetings (Bradley, 2007).

10.5 Audits

Three audits over a 6-month period had no impact on biosecurity compliance in eight poultry farms in Canada (Racicot et al., 2012a). So it is not always necessarily impactful. In order to be valid, the audit process must be relevant, objective, quantifiable, repeatable, and capable of suggesting changes to be made. Standards must be clearly identified. These standards must reconcile current practices with protocols already defined, evaluated, and published. The data must collectively be the subject of regular reports showing the points assessed, the improvements or shortcomings identified, the corrective measures and their results. Finally, audits must be subject to independent evaluation to improve the process (Shaw and Costain, 1989; Smith, 1990). When these conditions are met, on-farm biosecurity audits have been shown to be very effective in aquaculture in New Zealand (Georgiades et al., 2016). In Belgium, an audit system, Biocheck.UGent, has been developed for swine and poultry and has made it possible to establish a link between the level of biosecurity and production performances (Gelaude et al., 2014; Rodrigues da Costa et al., 2019).

11 The economics of biosecurity

There is a paucity of scientific articles directly investigating the economic impact of biosecurity. An economic assessment of biosecurity in broiler breeders in 1987 indicated that a benefit–cost ratio of at least 3 should be expected for a farm considered at a 30% risk of being infected by an agent causing a severe disease (Gifford et al., 1987). Morris (1995) states that the primary purpose of any business should be to maximize return on investment over the long term. This is an important concept because a comprehensive biosecurity program does not necessarily offer a quick pay back. Yet, it is, or should be, an essential component of a farm owner's long-term strategy for success.

Biosecurity investments involve a mix of fixed and variable costs. Knowledge of costs would help inform cost-sharing programs related to animal disease mitigation efforts (Pudenz et al., 2019). In British Columbia, shortly after the H7N3 outbreak in 2004, these two types of costs were reported to represent about 4.5% of the cost of production for duck farms (JP Vaillancourt, pers. Comm.). However, in determining the economic impact of biosecurity measures, one must also consider the benefits in terms of reduced losses associated with infectious diseases, including the reduction in the use of antibiotics (Rojo-Gimeno et al., 2016). In pig production, a negative correlation between enzootic pneumonia, pleurisy, acute pleuropneumonia, and internal biosecurity underscores the importance of good biosecurity in reducing health problems (Pandolfi et al., 2018). In South Africa, financial modeling over a 3-year period of a farrow to finish pig farm of 122 sows estimated that, in the absence of disease, the implementation of biosecurity resulted in a 9.70% reduction in the total annual profit. In contrast, the study found that the implementation of biosecurity and its effective monitoring would prevent losses due to Africa Swine Fever with an impressive benefit–cost ratio of 29 (Fasina et al., 2012).

The perceptions of pastoralists on the cost of biosecurity are important for the adoption of these measures. If the perceived costs are excessive, producers may prefer not to apply the measures. This is in accordance with Casal et al. (2007), who found that perceptions about biosecurity measures and their use were interrelated. Other studies such as Fraser et al. (2010), Valeeva et al. (2011), and Toma et al. (2013) also found that the costs and benefits of biosecurity contribute to the adoption of biosecurity measures (Niemi et al., 2016). But the likelihood of adopting biosecurity increases inelastically as perceived costs decrease. This means that the relative change (percent) in biosecurity adoption is smaller than the relative change in assumed cost. On the other hand, if the grower receives information about a low-cost biosecurity technology that is available, it increases the rate of biosecurity adoption.

12 Conclusion and future trends

We essentially know the risk factors associated with infectious disease transmission in poultry. Biosecurity measures have been developed over the years. Two main challenges require research for on-farm measures: (a) finding ways to make specific measures easier, faster, and cheaper to perform; (b) how to increase compliance. Any progress in point (a) will favor point (b). But this second point has been associated with many factors, namely, lack of knowledge; economic constraints; lack of training, communication, incentives, time; difficulty in applying the requested measures; lack of audits; lack of consistency in the information available; beliefs, attitudes, perceptions, education, experience, and personality traits of farm workers. It is essential to consider several of these factors at the same time if one wants to have a significant impact on biosecurity compliance.

Finally, research is needed regarding regional biosecurity measures, such as the management of poultry traffic, zone raising (regional downtime), and zoning.

13 Where to look for further information

Good introductions to the subject for non-specialists:

- Dewulf, J. and Van Immerseel, F. (Eds). (2019). *Biosecurity in Animal Production and Veterinary Medicine*. CABI. 523 pp.
- Owen, R. L. (2011). *A Practical Guide for Managing Risk in Poultry Production* (No. V480 OWEp). 276 pp.
- Racicot, M., Venne D., Durivage A. and Vaillancourt J.-P. (2011). Description of 44 biosecurity errors while entering and exiting poultry barns based on video surveillance in Quebec, Canada. *Prev. Vet. Med.* 100, 193–199.
- Conan, A., Goutard, F. L., Sorn, S. and Vong, S. (2012). Biosecurity measures for backyard poultry in developing countries: a systematic review. *BMC Veterinary Research* 8(1), 1–10.
- Scott, A. B., Singh, M., Groves, P., Hernandez-Jover, M., Barnes, B., Glass, K., Moloney, B., Black, A. and Toribio, J. A. (2018). Biosecurity practices on Australian commercial layer and meat chicken farms: Performance and perceptions of farmers. *PLoS ONE* 13(4), e0195582.

Key research centers readers can investigate, for example, for possible collaboration as well as to keep up with research trends:

1. Prof. Dr. Jeroen Dewulf: https://biocheck.ugent.be/en:
    Veterinary Epidemiology Unit, Department of Reproduction, Obstetrics and Herd Health.

Faculty of Veterinary Medicine; Ghent University, Belgium: https://www.ugent.be/di/vvb/en/research/research-epidemiology.
2. Prof. Jean-Pierre Vaillancourt: Department of Clinical Sciences; Faculty of Veterinary Medicine, Université de Montréal, Canada. Jean-pierre.vaillancourt@umontreal.ca.
3. Jean-Luc Guérin. Poultry Biosecurity Chair, École Nationale Vétérinaire de Toulouse; jl.guerin@envt.fr: http://www.envt.fr/content/recherche-publication-de-la-chaire-de -biosécurité-aviaire.
4. Alberto Oscar Allepuz Palau: Alberto.Allepuz@uab.cat: Centre de Recerca en Sanitat Animal IRTA-CReSA; University of Barcelona, Spain.
5. Armin R. W. Elbers; armin.elbers@wur.nl: Department of Epidemiology, Bioinformatics and Animal Models, Wageningen Bioveterinary Research, The Netherlands.

14 References

Allen, V. and Newell, D. (2005). *Evidence for the Effectiveness of Biosecurity to Exclude Campylobacter from Poultry Flocks*. Commissioned Project MS0004. Food Standards Agency Report. Available at: https://pdfs.semanticscholar.org/4236/1dd1fb80d51 742c54629ee46cb315ef936a3.pdf?_ga=2.63452588.287620099.1573885102 -1033270699.1521297455 (accessed 22 March 2021).

Amass, S. F. (1999). Biosecurity considerations for pork production units, *Swine Health and Production* 7(5), 12.

Amass, S. F., Vyverberg, B. D., Ragland, D., Dowell, C. A., Anderson, C. D., Stover, J. H. and Beaudry, D. J. (2000). Evaluating the efficacy of boot baths in biosecurity protocols, *Journal of Swine Health and Production* 8(4), 169-173.

Amass, S. F., Arighi, M., Kinyon, J. M., Hoffman, L. J., Schneider, J. L. and Draper, D. K. (2006). Effectiveness of using a mat filled with a peroxygen disinfectant to minimize shoe sole contamination in a veterinary hospital, *Journal of the American Veterinary Medical Association* 228(9), 1391-1396.

Anon. (2018). National on-farm avian biosecurity standards. Available at: https:// inspection.canada.ca/animal-health/terrestrial-animals/biosecurity/standards-and -principles/national-avian-on-farm-biosecurity-standard/eng/1528732756921 /1528732872665?chap=0 (accessed 18 April 2022).

Arsenault, J., Letellier, A., Quessy, S., Normand, V. and Boulianne, M. (2007). Prevalence and risk factors for *Salmonella* spp. and *Campylobacter* spp. caecal colonization in broiler chicken and turkey flocks slaughtered in Quebec, Canada, *Preventive Veterinary Medicine* 81(4), 250-264.

Axtell, R. C. (1999). Poultry integrated pest management: status and future, *Integrated Pest Management Reviews* 4(1), 53-73.

Bandura, A. and Cervone, D. (1983). Self-evaluative and self-efficacy mechanisms governing the motivational effects of goal systems, *Journal of Personality and Social Psychology* 45(5), 1017-1028. doi: 10.1037/0022-3514.45.5.1017.

Barcelo, M. and Marco, E. (1998). On farm biosecurity. *Proceedings of the 15th Intl Pig. Veterinary Society Congress*. Nottingham University Press, Birmingham, England.

Barrington, G. M., Allen, A. J., Parish, S. M. and Tibary, A. (2006). Biosecurity and biocontainment in alpaca operations, *Small Ruminant Research* 61(2), 217–225. doi: 10.1016/j.smallrumres.2005.07.011.

Bates, C., Hiett, K. L. and Stern, N. J. (2004). Relationship of campylobacter isolated from poultry and from darkling beetles in New Zealand, *Avian Diseases* 48(1), 138–147.

Battersby, T., Walsh, D., Whyte, P. and Bolton, D. (2017). Evaluating and improving terminal hygiene practices on broiler farms to prevent Campylobacter cross-contamination between flocks, *Food Microbiology* 64, 1–6. doi: 10.1016/j.fm.2016.11.018.

Bennett, B. (2017). The importance of biosecurity in the modern day hatchery, *International Hatchery Practice* 31, 21–23.

Bestman, M., de Jong, W., Wagenaar, J. and Weerts, T. (2018). Presence of avian influenza risk birds in and around poultry free-range areas in relation to range vegetation and openness of surrounding landscape, *Agroforestry Systems* 92(4), 1001–1008. doi: 10.1007/s10457-017-0117-2.

Blondel, V., Huard, G., Vaillancourt, J. P. and Racicot, M. (2018). Base du nettoyage et de la désinfection dans les exploitations agricoles. Available at: https://www.agrireseau.net/documents/98011/bases-du-nettoyage-et-de-la-desinfection-dans-les-exploitations-agricoles (accessed 7 October 2021).

Boender, G. J., Meester, R., Gies, E. and De Jong, M. C. (2007). The local threshold for geographical spread of infectious diseases between farms, *Preventive Veterinary Medicine* 82(1–2), 90–101. doi: 10.1016/j.prevetmed.2007.05.016.

Böhm, R. (1998). Disinfection and hygiene in the veterinary field and disinfection of animal houses and transport vehicles, *International Biodeterioration and Biodegradation* 41(3–4), 217–224.

Boklund, A., Alban, L., Mortensen, S. and Houe, H. (2004). Biosecurity in 116 Danish fattening swineherds: descriptive results and factor analysis, *Preventive Veterinary Medicine* 66(1–4), 49–62. doi: 10.1016/j.prevetmed.2004.08.004.

Bonhotal, J., Schwarz, M. and Rynk, R. (2014). Composting animal mortalities. Cornell Waste Management Institute, 1–23. Available at: https://hdl.handle.net/1813/37369.

Bouwstra, R., Gonzales, J. L., de Wit, S., Stahl, J., Fouchier, R. A. M. and Elbers, A. R. W. (2017). Risk for low pathogenicity avian influenza virus on poultry farms, the Netherlands, 2007–2013, *Emerging Infectious Diseases* 23(9), 1510–1516. doi: 10.3201/eid2309.170276.

Bradley, F. A. (2007). Biosecurity: educational programs, *Journal of Applied Poultry Research* 16(1), 77–81. doi: 10.1093/japr/16.1.77.

Brglez, B. (2003). *Disposal of Poultry Carcasses in Catastrophic Avian Influenza Outbreaks*. doi: 10.17615/vxd0-mv88.

Brody, S. N. (1974). The disease of the soul: leprosy in medieval literature (1st ed.). Ithaca: Cornell University Press.

Buckalew, L. W. and Sallis, R. E. (1986). Patient compliance and medication perception, *Journal of Clinical Psychology* 42(1), 49–53. doi: 10.1002/1097-4679(198601)42:1< 49::AID-JCLP2270420107>3.0.CO;2-F.

Butcher, G. D. and Miles, R. D. (1995). *Minimizing Microbial Contamination in Feed Mills Producing Poultry Feed*. Veterinary Medicine-Large Animal Clinical Sciences Department, Florida Cooperative Extension Service, Institute of Food and

Agricultural Sciences and University of Florida. Available at: http://www.nutritime .com.br/arquivos_internos/artigos/Artigo78_VM054002.pdf.

Casal, J., De Manuel, A., Mateu, E. and Martín, M. (2007). Biosecurity measures on swine farms in Spain: perceptions by farmers and their relationship to current on-farm measures, *Preventive Veterinary Medicine* 82(1–2), 138–150. doi: 10.1016/j. prevetmed.2007.05.012.

Cassar, J. R., Bright, L. M., Patterson, P. H., Mills, E. W. and Demirci, A. (2020). The efficacy of pulsed ultraviolet light processing for table and hatching eggs, *Poultry Science* 100(3), 100923.

Cerf, O., Carpentier, B. and Sanders, P. (2010). Tests for determining in-use concentrations of antibiotics and disinfectants are based on entirely different concepts: "Resistance" has different meanings, *International Journal of Food Microbiology* 136(3), 247–254.

Chaber, A. L. and Saegerman, C. (2017). Biosecurity measures applied in the United Arab Emirates – a comparative study Between livestock and wildlife sectors, *Transboundary and Emerging Diseases* 64(4), 1184–1190. doi: 10.1111/tbed.12488.

Chaudhry, M., Ahmad, M., Rashid, H. B., Sultan, B., Chaudhry, H. R., Riaz, A. and Shaheen, M. S. (2017). Prospective study of avian influenza H9 infection in commercial poultry farms of Punjab Province and Islamabad Capital Territory, Pakistan, *Tropical Animal Health and Production* 49(1), 213–220. doi: 10.1007/s11250-016-1159-6. PMID: 27761776; PMCID: PMC7088531.

Chen, S. J., Hung, M. C., Huang, K. L. and Hwang, W. I. (2004). Emission of heavy metals from animal carcass incinerators in Taiwan, *Chemosphere* 55(9), 1197–1205.

Chin, R. P., García, M., Corsiglia, C., Riblet, S., Crespo, R., Shivaprasad, H. L., Rodríguez-Avila, A., Woolcock, P. R. and França, M. (2009). Intervention strategies for laryngotracheitis: impact of extended downtime and enhanced biosecurity auditing, *Avian Diseases* 53(4), 574–577. doi: 10.1637/8873-041309-Reg.1.

Cochrane, R. A. (2016). Feed mill biosecurity plans: a systematic approach to prevent biological pathogens in swine feed, *Journal of Swine Health and Production* 24(3), 154–164.

Colombo, C., Giger, H., Grote, J., Deplazes, C., Pletscher, W., Lüthi, R. and Ruef, C. (2002). Impact of teaching interventions on nurse compliance with hand disinfection, *Journal of Hospital Infection* 51(1), 69–72. doi: 10.1053/jhin.2002.1198.

Conly, J. M., Hill, S., Ross, J., Lertzman, J. and Louie, T. J. (1989). Handwashing practices in an intensive care unit: the effects of an educational program and its relationship to infection rates, *American Journal of Infection Control* 17(6), 330–339. doi: 10.1016/0196-6553(89)90002-3.

Corrigan, R. M. (2006). Overview of rodent control for commercial pork operation. Available at: https://porkgateway.org/resource/an-overview-of-rodent-control-for -commercial-pork-production-operations/.

Course, C. E., Boerlin, P., Slavic, D., Vaillancourt, J. P. and Guerin, M. T. (2021). Factors associated with *Salmonella enterica* and *Escherichia coli* during downtime in commercial broiler chicken barns in Ontario, *Poultry Science* 100(5), 101065.

Crippen, T. L. and Sheffield, C. (2006). External surface disinfection of the lesser mealworm (Coleoptera: Tenebrionidae), *Journal of Medical Entomology* 43(5), 916–923.

Curtis, P. E., Ollerhead, G. E. and Ellis, C. E. (1980). *Pasteurella multocida* infection of poultry farm rats, *Veterinary Record* 107(14), 326–327.

Curtis, P. E. and Ollerhead, G. E. (1982). *Pasteurella multocida* infection of cats on poultry farms, *Veterinary Record* 110(1), 13-14.

Davies, R. H. and Wray, C. (1995). Observations on disinfection regimens used on *Salmonella enteritidis* infected poultry units, *Poultry Science* 74(4), 638-647.

Davison, S. A., Dunn, P. A., Henzler, D. J., Knabel, S. J., Patterson, P. H. and Schwartz, J. H. (1997). *Preharvest HACCP in the Table Egg Industry*. PA State: The Pennsylvania State University, 1-36.

Dee, S., Deen, J., Rossow, K., Wiese, C., Otake, S., Joo, H. S. and Pijoan, C. (2002). Mechanical transmission of porcine reproductive and respiratory syndrome virus throughout a coordinated sequence of events during cold weather, *Canadian Journal of Veterinary Research* 66(4), 232-239.

Dee, S., Deen, J. and Pijoan, C. (2004). Evaluation of 4 intervention strategies to prevent the mechanical transmission of porcine reproductive and respiratory syndrome virus, *Canadian Journal of Veterinary Research* 68(1), 19-26.

Delabbio, J. L., Johnson, G. R., Murphy, B. R., Hallerman, E., Woart, A. and McMullin, S. L. (2005). Fish disease and biosecurity: attitudes, beliefs, and perceptions of managers and owners of commercial finfish recirculating facilities in the United States and Canada, *Journal of Aquatic Animal Health* 17(2), 153-159. doi: 10.1577/H04-005.1.

Delpont, M., Racicot, M., Durivage, A., Fornili, L., Guerin, J. L., Vaillancourt, J. P. and Paul, M. C. (2020). Determinants of biosecurity practices in French duck farms after a H5N8 Highly Pathogenic Avian Influenza epidemic: the effect of farmer knowledge, attitudes and personality traits, *Transboundary and Emerging Diseases* 68(1), 51-61.

Dorea, F. C., Berghaus, R., Hofacre, C. and Cole, D. J. (2010). Survey of biosecurity protocols and practices adopted by growers on commercial poultry farms in Georgia, USA, *Avian Diseases* 54(3), 1007-1015.

Dunowska, M., Morley, P. S., Patterson, G., Hyatt, D. R. and Van Metre, D. C. (2006). Evaluation of the efficacy of a peroxygen disinfectant-filled footmat for reduction of bacterial load on footwear in a large animal hospital setting, *Journal of the American Veterinary Medical Association* 228(12), 1935-1939.

Duvauchelle, A., Huneau-Salaün, A., Balaine, L., Rose, N. and Michel, V. (2013). Risk factors for the introduction of avian influenza virus in breeder duck flocks during the first 24 weeks of laying, *Avian Pathology* 42(5), 447-456. doi: 10.1080/03079457.2013.823145.

East, I., Kite, V., Daniels, P. and Garner, G. (2006). A cross-sectional survey of Australian chicken farms to identify risk factors associated with seropositivity to Newcastle-disease virus, *Preventive Veterinary Medicine* 77(3-4), 199-214. doi: 10.1016/j.prevetmed.2006.07.004.

Ebeling, W. (1975). *Urban Entomology*. Los Angeles: University of California.

Elfadil, A. A., Vaillancourt, J. P. and Meek, A. H. (1996). Management risk factors associated with cellulitis in broiler chickens in Southern Ontario: a retrospective study, *Avian Diseases* 40(3), 699-706.

Ellis, D. B. (2001). Carcass disposal issues in recent disasters, accepted methods, and suggested plan to mitigate future events. Available at: https://digital.library.txstate.edu/handle/10877/3502 (accessed 18 April 2022).

England, J. J. (2002). Biosecurity: safeguarding your veterinarian:client:patient relationship, *Veterinary Clinics of North America. Food Animal Practice* 18(3), 373-8, v. doi: 10.1016/S0749-0720(02)00033-6.

Ernst, A. R. (2004). *Hatching Egg Sanitation: The Key Step in Successful Storage and Production*. University of California, Division of Agriculture and National Resources: Publication 8120.

Fasina, F. O., Lazarus, D. D., Spencer, B. T., Makinde, A. A. and Bastos, A. D. S. (2012). Cost implications of African swine fever in smallholder farrow-to-finish units: economic benefits of disease prevention through biosecurity, *Transboundary and Emerging Diseases* 59, 244-255.

Fernandez, D., et al. (1994). Farm location as a determinant to production performance in turkeys. Annual Meeting of the American Association of Avian Pathologists, Annual Convention AVMA, San Francisco.

Fraser, R. W., Williams, N. T., Powell, L. F. and Cook, A. J. C. (2010). Reducing campylobacter and salmonella infection: two studies of the economic cost and attitude to adoption of on-farm biosecurity measures, *Zoonoses and Public Health* 57(7-8), e109-e115.

Gelaude, P., Schlepers, M., Verlinden, M., Laanen, M. and Dewulf, J. (2014). Biocheck. UGent: a quantitative tool to measure biosecurity at broiler farms and the relationship with technical performances and antimicrobial use, *Poultry Science* 93(11), 2740-2751.

Georgiades, E., Fraser, R. and Jones, B. (2016). Options to strengthen on-farm biosecurity management for commercial and non-commercial aquaculture. Aquaculture Unit. Technical Paper No: 2016/47.

Gifford, D. H., Shane, S. M., Hugh-Jones, M. and Weigler, B. J. (1987). Evaluation of biosecurity in broiler breeders, *Avian Diseases* 31(2), 339-344. doi: 10.2307/1590882.

Glanville, T. D., Ahn, H. K., Richard, T. L., Harmon, J. D., Reynolds, D. L. and Akinc, S. (2006). Environmental impacts of emergency livestock mortality composting-leachate release and soil contamination. 2006 ASAE Annual Meeting (p. 1). American Society of Agricultural and Biological Engineers.

Graham, J. P., Price, L. B., Evans, S. L., Graczyk, T. K. and Silbergeld, E. K. (2009). Antibiotic resistant enterococci and staphylococci isolated from flies collected near confined poultry feeding operations, *Science of the Total Environment* 407(8), 2701-2710.

Guinat, C., Comin, A., Kratzer, G., Durand, B., Delesalle, L., Delpont, M., Guérin, J. L. and Paul, M. C. (2020). Biosecurity risk factors for highly pathogenic avian influenza (H5N8) virus infection in duck farms, France, *Transboundary and Emerging Diseases* 67(6), 2961-2970. doi: 10.1111/tbed.13672.

Gunn, G. J., Heffernan, C., Hall, M., McLeod, A. and Hovi, M. (2008). Measuring and comparing constraints to improved biosecurity amongst GB farmers, veterinarians and the auxiliary industries, *Preventive Veterinary Medicine* 84(3-4), 310-323. doi: 10.1016/j.prevetmed.2007.12.003.

Hald, B., Skovgard, H., Bang, D. D., Pedersen, K., Dybdahl, J., Jespersen, J. B. and Madsen, M. (2004). Flies and campylobacter infection of broiler flocks, *Emerging Infectious Diseases* 10(8), 1490-1492.

Halvorson, D. A. (2002). The control of H5 or H7 mildly pathogenic avian influenza: a role for inactivated vaccine. *Avian Pathology* 31(1):5-12. doi: 10.1080/03079450120106570. PMID: 12430550.

Hauck, R., Crossley, B., Rejmanek, D., Zhou, H. and Gallardo, R. A. (2017). Persistence of highly pathogenic and low pathogenic avian influenza viruses in footbaths and poultry manure. *Avian Diseases* 61(1), 64-69. doi: 10.1637/11495-091916-Reg.

Haynes, R. B., Taylor, D. W. and Sackett, D. L. (Eds) (1979). *Compliance in Health Care*. Baltimore: Johns Hopkins University Press.

Henzler, D. J. and Opitz, H. M. (1992). The role of mice in the epizootiology of *Salmonella enteritidis* infection on chicken layer farms, *Avian Diseases* 36(3), 625–631.

Jardine, C. G. and Hrudey, S. E. (1997). Mixed messages in risk communication, *Risk Analysis* 17(4), 489–498. doi: 10.1111/j.1539-6924.1997.tb00889.x.

Jeong, J., Kang, H. M., Lee, E. K., Song, B. M., Kwon, Y. K., Kim, H. R., Choi, K. S., Kim, J. Y., Lee, H. J., Moon, O. K., Jeong, W., Choi, J., Baek, J. H., Joo, Y. S., Park, Y. H., Lee, H. S. and Lee, Y. J. (2014). Highly pathogenic avian influenza virus (H5N8) in domestic poultry and its relationship with migratory birds in South Korea during 2014, *Veterinary Microbiology* 173(3-4), 249–257.

Johnson, Y. J., Gedamu, N., Colby, M. M., Myint, M. S., Steele, S. E., Salem, M. and Tablante, N. L. (2005). Wind-borne transmission of infectious laryngotracheitis between commercial poultry operations, *International Journal of Poultry Science* 4(5), 263–267. doi: 10.3923/ijps.2005.263.267.

Jones, F. T. (2011). A review of practical Salmonella control measures in animal feed, *Journal of Applied Poultry Research* 20(1), 102–113.

Pinto, C. J. and Urcelay, V. S. (2003). Biosecurity practices on intensive pig production systems in Chile, *Preventive Veterinary Medicine* 59(3), 139–145. doi: 10.1016/S0167-5877(03)00074-6.

Kalbasi, A., Mukhtar, S., Hawkins, S. E. and Auvermann, B. W. (2005). Carcass composting for management of farm mortalities: a review, *Compost Science and Utilization* 13(3), 180–193.

Keener, H. M., Elwell, D. L. and Monnin, M. J. (2000). Procedures and equations for sizing of structures and windrows for composting animal mortalities, *Applied Engineering in Agriculture* 16(6), 681–692.

Kim, J. H. and Kim, K. S. (2010). Hatchery hygiene evaluation by microbiological examination of hatchery samples, *Poultry Science* 89(7), 1389–1398.

Kim, W. H., An, J. U., Kim, J., Moon, O. K., Bae, S. H., Bender, J. B. and Cho, S. (2018). Risk factors associated with highly pathogenic avian influenza subtype H5N8 outbreaks on broiler duck farms in South Korea, *Transboundary and Emerging Diseases* 65(5), 1329–1338.

King, M. A., Seekins, B., Hutchinson, M. and MacDonald, G. (2005). Observations of static pile composting of large animal carcasses using different media. Symposium on Composting Animal Mortalities and Slaughterhouse Residuals in South Portland, ME.

Korbel, R., Gerlach, H., Bisgaard, M. and Hafez, H. M. (1992). Further investigations on *Pasteurella multocida* infections in feral birds injured by cats, *Zentralblatt Fur Veterinarmedizin. Reihe B. Journal of Veterinary Medicine. Series B* 39(1), 10–18.

Kuney, D. R. and Jeffrey, J. S. (2002). Cleaning and disinfecting poultry facilities. In: Bell, D. D., Weaver, W. D. (Eds) *Commercial Chicken Meat and Egg Production*, 557–564. Boston: Springer. doi: 10.1007/978-1-4615-0811-3_29.

Langsrud, S., Møretrø, T. and Sundheim, G. (2003). Characterization of *Serratia marcescens* surviving in disinfecting footbaths, *Journal of Applied Microbiology* 95(1), 186–195. doi: 10.1046/j.1365-2672.2003.01968.x.

Larsen, A. F. (2009). Semi-subsistence producers and biosecurity in the Slovenian alps, *Sociologia Ruralis* 49(4), 330–343. doi: 10.1111/j.1467-9523.2009.00481.x.

Lee, H. J., Jeong, J. Y., Jeong, O. M., Youn, S. Y., Kim, J. H., Kim, D. W., Yoon, J. U., Kwon, Y. K. and Kang, M. S. (2020). Impact of Dermanyssus gallinae infestation on persistent outbreaks of fowl typhoid in commercial layer chicken farms, *Poultry Science* 99(12), 6533-6541.

Li, X., Bethune, L. A., Jia, Y., Lovell, R. A., Proescholdt, T. A., Benz, S. A., Schell, T. C., Kaplan, G. and McChesney, D. G. (2012). Surveillance of *Salmonella* prevalence in animal feeds and characterization of the *Salmonella* isolates by serotyping and antimicrobial susceptibility, *Foodborne Pathogens and Disease* 9(8), 692-698. doi: 10.1089/fpd.2011.1083.

Losinger, W. C., Bush, E. J., Hill, G. W., Smith, M. A., Garber, L. P., Rodriguez, J. M. and Kane, G. (1998). Design and implementation of the United States National Animal Health Monitoring System 1995 National Swine Study, *Preventive Veterinary Medicine* 34(2-3), 147-159. doi: 10.1016/s0167-5877(97)00076-7.

Lotz, J. M. (1997). Special topic review: viruses, biosecurity and specific pathogen-free stocks in shrimp aquaculture, *World Journal of Microbiology and Biotechnology* 13, 9.

Mauldin, J. M. (1983). *Hatchery and Breeder Flock Sanitation Guide*. Bulletin-Cooperative Extension Service. Available at: agris.fao.org.

Mauldin, J. M. (2002). Hatchery planning, design, and construction. In: Bell, D. D., Weaver, W. D. and North, M. O. (Eds) *Commercial Chicken Meat and Egg Production*, 661-683. Boston: Springer.

Mauldin, J. M. and MacKinnon, I. R. (2009). Hatchery ventilation and environmental control, *Avian Biology Research* 2(1-2), 87-91.

McQuiston, J. H., Garber, L. P., Porter-Spalding, B. A., Hahn, J. W., Pierson, F. W., Wainwright, S. H., Senne, D. A., Brignole, T. J., Akey, B. L. and Holt, T. J. (2005). Evaluation of risk factors for the spread of low pathogenicity H7N2 avian influenza virus among commercial poultry farms, *Journal of the American Veterinary Medical Association* 226(5), 767-772.

Millman, C., Christley, R., Rigby, D., Dennis, D., O'Brien, S. J. and Williams, N. (2017). "Catch 22": biosecurity awareness, interpretation and practice amongst poultry catchers, *Preventive Veterinary Medicine* 141, 22-32.

Moore, D. A., Merryman, M. L., Hartman, M. L. and Klingborg, D. J. (2008). Comparison of published recommendations regarding biosecurity practices for various production animal species and classes, *Journal of the American Veterinary Medical Association* 233(2), 249-256. doi: 10.2460/javma.233.2.249.

Morley, P. S., Morris, S. N., Hyatt, D. R. and Van Metre, D. C. (2005). Evaluation of the efficacy of disinfectant footbaths as used in veterinary hospitals, *Journal of the American Veterinary Medical Association* 226(12), 2053-2058.

Moro, C. V., De Luna, C. J., Tod, A., Guy, J. H., Sparagano, O. A. and Zenner, L. (2009). The poultry red mite (Dermanyssus gallinae): a potential vector of pathogenic agents. In: Sparagano, O. A. (Ed.) *Control of Poultry Mites (Dermanyssus)*, 93-104. Dordrecht: Springer.

Morris, M. P. (1995). Economic considerations in prevention and control of poultry disease. In: *Biosecurity in the Poultry Industry*, Shane, S. M., Halvorson, D., Hill, D., Villegas, P. and Wages, D. (Eds), 4-16. Kennett Square: American Association of Avian Pathologists.

Mukhtar, S., Kalbasi, A. and Ahmed, A. (2004). *Composting: A Comprehensive Review*. Kansas State University, National Agricultural Biosecurity Center. Available at: https://krex.k-state.edu/dspace/bitstream/handle/2097/662/Chapter3.pdf?sequence=16&isAllowed=y.

Nespeca, R., Vaillancourt, J. P. and Morrow, W. E. (1997). Validation of a poultry biosecurity survey, *Preventive Veterinary Medicine* 31(1–2), 73–86.

Niemi, J. K., Sahlström, L., Kyyrö, J., Lyytikäinen, T. and Sinisalo, A. (2016). Farm characteristics and perceptions regarding costs contribute to the adoption of biosecurity in Finnish pig and cattle farms, *Review of Agricultural, Food and Environmental Studies* 97(4), 215–223. doi: 10.1007/s41130-016-0022-5.

Nishiguchi, A., Kobayashi, S., Yamamoto, T., Ouchi, Y., Sugizaki, T. and Tsutsui, T. (2007). Risk factors for the introduction of avian influenza virus into commercial layer chicken farms During the outbreaks caused by a low-pathogenic H5N2 virus in Japan in 2005, *Zoonoses and Public Health* 54(9–10), 337–343. doi: 10.1111/j.1863-2378.2007.01074.x.

O'Boyle, C. A., Henly, S. J. and Larson, E. (2001). Understanding adherence to hand hygiene recommendations: the theory of planned behavior, *American Journal of Infection Control* 29(6), 352–360. doi: 10.1067/mic.2001.18405.

Ojewole, A. O. (2011). Biblical mandates for sustainable sanitation, *Millennium Development Goals (MdGS) as Instrument for Development in Africa*, 598–613. Available at: https://d1wqtxts1xzle7.cloudfront.net/49544347/How_green_are_hotels_in_Accra_Environmen20161012-27178-u5ze10-with-cover-page-v2.pdf?Expires=1650324230&Signature=JFBCUcC8IbV5I4WMUpjOQBc-xpAsgUu0WZKcq-ZnnZ7pq2Ua~ccfwUSN-6YyQkCAoTIxr1h0qks8EVWkpaA3m315SoPjCoFUEGFka9-3QYp9wC7rRDRSAX7a~g9EgMH253tYuJW5Z67hMcLH5mNQwdLwmCFQmwAaLzzvk4c-xh5V6I-1Y576Klzv8SkCS0giTlCCqRfmb-Tv5hmh9JoKz9K0C7qRmHu0XlmMf31j6NecmUm-Xhh0PwelFFxkoGL~BzwkZ5TQ~Upyb7RefKAOjze65X2XdCCuri-14XVRvBK~n-lTgh3FoX-73onLAx0QfXr7ACdxvSgotDmDxYvmBQ__&Key-Pair-Id=APKAJLOHF5GGSLRBV4ZA#page=609.

Owen, R. L. and Lawlor, J. (2012). A novel approach to foot dipping. Available at: https://fr.slideserve.com/bernad/a-novel-approach-to-foot-dipping.

Pagès-Manté, A., Torrents, D., Maldonado, J. and Saubi, N. (2004). Dogs as potential carriers of infectious bursal disease virus, *Avian Pathology* 33(2), 205–209.

Pandolfi, F., Edwards, S. A., Maes, D. and Kyriazakis, I. (2018). Connecting different data sources to assess the interconnections between biosecurity, health, welfare, and performance in commercial pig farms in Great Britain, *Frontiers in Veterinary Science* 5, 41. doi: 10.3389/fvets.2018.00041.

Payne, J. B., Kroger, E. C. and Watkins, S. E. (2005). Evaluation of disinfectant efficacy when applied to the floor of poultry grow-out facilities, *Journal of Applied Poultry Research* 14(2), 322–329.

Pervin, L. A. and John, O. P. (2001). *Personality: Theory and Research* (8th edn.). New York: John Wiley & Sons Inc.

Pudenz, C. C., Schulz, L. L. and Tonsor, G. T. (2019). Adoption of secure pork supply plan biosecurity by US Swine producers, *Frontiers in Veterinary Science* 6, 146.

Rabie, A. J., McLaren, I. M., Breslin, M. F., Sayers, R. and Davies, R. H. (2015). Assessment of anti-Salmonella activity of boot dip samples, *Avian Pathology* 44(2), 129–134.

Racicot, M., Venne, D., Durivage, A. and Vaillancourt, J. P. (2011). Description of 44 biosecurity errors while entering and exiting poultry barns based on video surveillance in Quebec, Canada, *Preventive Veterinary Medicine* 100(3–4), 193–199. doi: 10.1016/j.prevetmed.2011.04.011.

Racicot, M., Venne, D., Durivage, A. and Vaillancourt, J. P. (2012a). Evaluation of strategies to enhance biosecurity compliance on poultry farms in Québec: effect of audits and cameras, *Preventive Veterinary Medicine* 103(2-3), 208-218. doi: 10.1016/j. prevetmed.2011.08.004.

Racicot, M., Venne, D., Durivage, A. and Vaillancourt, J. P. (2012b). Evaluation of the relationship between personality traits, experience, education and biosecurity compliance on poultry farms in Québec, Canada, *Preventive Veterinary Medicine* 103(2-3), 201-207. doi: 10.1016/j.prevetmed.2011.08.011.

Racicot, M., Comeau, G., Tremblay, A., Quessy, S., Cereno, T., Charron-Langlois, M., Venne, D., Hébert, G., Vaillancourt, J. P., Fravalo, P., Ouckama, R., Mitevski, D., Guerin, M. T., Agunos, A., DeWinter, L., Catford, A., Mackay, A. and Gaucher, M. L. (2020). Identification and selection of food safety-related risk factors to be included in the Canadian Food Inspection Agency's Establishment-based Risk Assessment model for Hatcheries, *Zoonoses and Public Health* 67(1), 14-24.

Refrégier-Petton, J., Rose, N., Denis, M. and Salvat, G. (2001). Risk factors for Campylobacter spp. contamination in French broiler-chicken flocks at the end of the rearing period, *Preventive Veterinary Medicine* 50(1-2), 89-100. doi: 10.1016/s0167-5877(01)00220-3.

Ribbens, S., Dewulf, J., Koenen, F., Mintiens, K., De Sadeleer, L., de Kruif, A. and Maes, D. (2008). A survey on biosecurity and management practices in Belgian pig herds, *Preventive Veterinary Medicine* 83(3-4), 228-241. doi: 10.1016/j. prevetmed.2007.07.009.

Ricke, S. C., Richardson, K. and Dittoe, D. K. (2019). Formaldehydes in feed and their potential interaction with the poultry gastrointestinal tract microbial community-a review, *Frontiers in Veterinary Science* 6, 188. doi: 10.3389/fvets.2019.00188.

Roche, A. J., Cox, N. A., Richardson, L. J., Buhr, R. J., Cason, J. A., Fairchild, B. D. and Hinkle, N. C. (2009). Transmission of *Salmonella* to broilers by contaminated larval and adult lesser mealworms, *Alphitobius diaperinus* (Coleoptera: Tenebrionidae), *Poultry Science* 88(1), 44-48.

Rodgers, J. D., McCullagh, J. J., McNamee, P. T., Smyth, J. A. and Ball, H. J. (2001). An investigation into the efficacy of hatchery disinfectants against strains of Staphylococcus aureus associated with the poultry industry, *Veterinary Microbiology* 82(2), 131-140.

Rodrigues da Costa, M., Gasa, J., Calderón Díaz, J. A., Postma, M., Dewulf, J., McCutcheon, G. and Manzanilla, E. G. (2019). Using the biocheck. UGent™ scoring tool in Irish farrow-to-finish pig farms: assessing biosecurity and its relation to productive performance, *Porcine Health Management* 5, 4.

Rojo-Gimeno, C., Postma, M., Dewulf, J., Hogeveen, H., Lauwers, L. and Wauters, E. (2016). Farm-economic analysis of reducing antimicrobial use whilst adopting improved management strategies on farrow-to-finish pig farms, *Preventive Veterinary Medicine* 129, 74-87. doi: 10.1016/j.prevetmed.2016.05.001.

Rose, N., Beaudeau, F., Drouin, P., Toux, J. Y., Rose, V. and Colin, P. (2000). Risk factors for Salmonella persistence after cleansing and disinfection in French broiler-chicken houses, *Preventive Veterinary Medicine* 44(1-2), 9-20. doi: 10.1016/S0167-5877(00)00100-8.

Roy, D. N. and Brown, A. W. A. (1954). *Entomology : (Medical & Veterinary) Including Insecticides & Insect & Rat Control.* Calcutta: Excelsior Press.

Sambeek, F. V., McMurray, B. L. and Page, R. K. (1995). Incidence of *Pasteurella multocida* in poultry house cats used for rodent control programs, *Avian Diseases* 39(1), 145-146.

Sanchez, J., Stryhn, H., Flensburg, M., Ersbøll, A. K. and Dohoo, I. (2005). Temporal and spatial analysis of the 1999 outbreak of acute clinical infectious bursal disease in broiler flocks in Denmark, *Preventive Veterinary Medicine* 71(3-4), 209-223. doi: 10.1016/j.prevetmed.2005.07.006.

Sander, J. E., Warbington, M. C. and Myers, L. M. (2002). Selected methods of animal carcass disposal, *Journal of the American Veterinary Medical Association* 220(7), 1003-1005.

Sanderson, M. W., Dargatz, D. A. and Garry, F. B. (2000). Biosecurity practices of beef cow-calf producers, *Journal of the American Veterinary Medical Association* 217(2), 185-189. doi: 10.2460/javma.2000.217.185.

Sawabe, K., Hoshino, K., Isawa, H., Sasaki, T., Hayashi, T., Tsuda, Y., Kurahashi, H., Tanabayashi, K., Hotta, A., Saito, T., Yamada, A. and Kobayashi, M. (2006). Detection and isolation of highly pathogenic H5N1 avian influenza A viruses from blow flies collected in the vicinity of an infected poultry farm in Kyoto, Japan, 2004, *American Journal of Tropical Medicine and Hygiene* 75(2), 327-332.

Schoof, H. F. (1959). How far do flies fly and what effect does flight pattern have on their control, *Pest Control* 27(4), 16-24.

Shaw, C. D. and Costain, D. W. (1989). Guidelines for medical audit: seven principles, *BMJ* 299(6697), 498-499.

Smith, T. (1990). Medical audit, *BMJ* 300(6717), 65-65.

Snow, L. C., Davies, R. H., Christiansen, K. H., Carrique-Mas, J. J., Cook, A. J. and Evans, S. J. (2010). Investigation of risk factors for Salmonella on commercial egg-laying farms in Great Britain, 2004-2005, *The Veterinary Record* 166(19), 579-586. doi: 10.1136/vr.b4801.

Springthorpe, S. (2000). La désinfection des surfaces et de l'équipement, *Journal of Cancer Dent. Assoc* 66, 558-560.

Steiner, J. J. (2020). *Disinfection of Hatching Eggs Using Low-Energy Electron Beam* (Doctoral dissertation, University of Zurich).

Strömberg, A., Broström, A., Dahlström, U. and Fridlund, B. (1999). Factors influencing patient compliance with therapeutic regimens in chronic heart failure: a critical incident technique analysis, *Heart and Lung* 28(5), 334-341. doi: 10.1053/hl.1999.v28.a99538.

Tablante, N. L. and Malone, G. W. (2006). Controlling avian influenza through in-house composting of depopulated flocks: sharing Delmarva's experience. Proceedings of the 2006 National Symposium on Carcass Disposal. Available at: https://www.animalmortmgmt.org/wp-content/uploads/2014/04/Controlling-AI-through-in-House-Composting-of-De-Populated-F.pdf.

Thermote, L. (2006). Effective hygiene within the hatchery, *International Hatchery Practice* 20(5), 18-21.

Tibballs, J. (1996). Teaching hospital medical staff to handwash, *The Medical Journal of Australia* 164(7), 395-398.

Timmermans, S. and Berg, M. (1997). Standardization in action: achieving local universality through medical protocols, *Social Studies of Science* 27(2), 273-305.

Toma, B., Vaillancourt, J.-P., Dufour, B., Eliot, M., Moutou, F., Marsh, W., Benet, J.-J., Sanaa, M. and Michel, P. (1999). *Dictionary of Veterinary Epidemiology*. Wiley.

Toma, L., Stott, A. W., Heffernan, C., Ringrose, S. and Gunn, G. J. (2013). Determinants of biosecurity behaviour of British cattle and sheep farmers-a behavioural economics analysis, *Preventive Veterinary Medicine* 108(4), 321-333.

Vaillancourt J-P. (1995). Infectious laryngo-tracheitis in broilers: a case-control study, *Annual Meeting of the American Association of Avian Pathologists*, July 11, 1995, Pittsburgh, Pennsylvania

Vaillancourt, J.-P. and Martinez, A. (2001). Relative importance of biosecurity measures: a delphi study. Annual Meeting of the American Association of Avian Pathologists, 138th Annual Convention AVMA, Boston.

Vaillancourt, J.-P. and Carver, D. K. (1998). Biosecurity: perception is not reality, *Poultry Digest* 57(6), 28–36.

Vaillancourt, J.-P., Delpont, M., Racicot, M., Paul, M. and Guérin, J.-L. (2018). Une perspective régionale de la biosécurité. Le nouveau praticien vétérinaire; 10/n°40156.

Valeeva, N. I., van Asseldonk, M. A. and Backus, G. B. (2011). Perceived risk and strategy efficacy as motivators of risk management strategy adoption to prevent animal diseases in pig farming, *Preventive Veterinary Medicine* 102(4), 284–295.

Van De Giessen, A. W., Tilburg, J. J. H. C., Ritmeester, W. S. and Van Der Plas, J. (1998). Reduction of campylobacter infections in broiler Flocks by application of hygiene measures, *Epidemiology and Infection* 121(1), 57–66.

Vandekerchove, D., De Herdt, P., Laevens, H. and Pasmans, F. (2004). Colibacillosis in caged layer hens: characteristics of the disease and the aetiological agent, *Avian Pathology* 33(2), 117–125. doi: 10.1080/03079450310001642149.

Verhagen, J. H., Fouchier, R. A. M. and Lewis, N. (2021). Highly pathogenic avian influenza viruses at the wild–domestic bird interface in Europe: future directions for research and surveillance, *Viruses* 13(2), 212.

Villa, B. and Velasco, A. (1994). Integrated pest management of the rat Rattus norvegicus in poultry farms, *Veterinaria México* 25(3), 247–249.

Volkova, V., Thornton, D., Hubbard, S. A., Magee, D., Cummings, T., Luna, L., Watson, J. and Wills, R. (2012). Factors associated with introduction of infectious laryngotracheitis virus on broiler farms during a localized outbreak, *Avian Diseases* 56(3), 521–528. doi: 10.1637/10046-122111-Reg.1.

Vroom, V. H. (1994). *Work and Motivation*. San Francisco, CA: John Wiley & Sons.

Watkins, S. and Venne, D. (2015). Water quality. In: *Manuel de pathologie aviaire*, Brugère-Picoux, J., Vaillancourt, J.-P., Bouzouaia, M., Shivaprasad, H. L. and Venne, D. (Eds), 560–569. Paris: AFAS.

Young, I., Rajić, A., Letellier, A., Cox, B., Leslie, M., Sanei, B. and McEwen, S. A. (2010). Knowledge and attitudes toward food safety and use of good production practices among Canadian broiler chicken producers, *Journal of Food Protection* 73(7), 1278–1287. doi: 10.4315/0362-028x-73.7.1278.

Zhang, Y. H., Li, C. S., Liu, C. C. and Chen, K. Z. (2013). Prevention of losses for hog farmers in China: insurance, on-farm biosecurity practices, and vaccination, *Research in Veterinary Science* 95(2), 819–824.

Chapter 4

Food safety management on farms producing beef

Peter Paulsen, Frans J. M. Smulders and Friederike Hilbert, University of Veterinary Medicine, Austria

1 Introduction

Experience has shown that the production of safe and wholesome food requires measures along the whole chain – from 'farm to fork' or 'from gate to plate'. The necessary measures and structures can be broadly separated into a set of 'prerequisites', often termed 'Good Farming Practice', and measures targeted against specific hazards to human or animal health ('interventions').

This chapter will describe the key elements of 'Good Farming Practice' comprising (1) a traditional 'hygiene' approach to control potential routes of contamination and carry-over, (2) the prerequisites needed to ensure that exposure of animals to hazards is minimized and (3) practices to ensure that animals are 'healthy' and not exposed to unnecessary stress and strain in the production cycle. This will include also management of the farm–environment interface (e.g. emissions from farms) and animal welfare issues.

Hazard-specific control measures in the sense of 'interventions' will be presented for only one selected biological hazard; details on all relevant biological and chemical hazards are discussed elsewhere in this volume.

http://dx.doi.org/10.19103/AS.2016.0008.07

2 Good farming practices and biosecurity for beef cattle farms

There is no universal definition of 'Good Farming Practices' (GFP), but it is obviously one facet of general 'Good Agricultural Practice' (GAP). According to FAO (2003), GAP 'applies available knowledge to addressing environmental, economic and social sustainability for on-farm production and post-production processes resulting in safe and healthy food and non-food agricultural products'. This means that in practice, legislation and standards addressing GFP often will address not only food safety, food security and animal welfare, but also environmental protection, rural development, common agricultural market and trade issues. The European Union gives a good example how complex the ramifications in legislation on GFP can be in large economic and political associations of nations (Bergschmidt et al., 2003).

The World Organisation of Animal Health (OIE) has developed a guide on GFP, with a focus on animal health (OIE Animal Production Food Safety Working Group, 2006). The guide addressed infectious and parasitic agents ('biological hazards'), chemical contaminants and residues ('chemical hazards') and foreign objects ('physical hazards') as well. The tools, namely codes of hygienic practice and Hazard Analysis and Critical Control Points (HACCP) or HACCP-based procedures, should be applied in order to implement preventive actions in eight areas of primary production (Table 1). The guide was not intended for direct use by farmers, but rather as an instrument for the competent authorities to assist farmers to assume their responsibility in the food chain, for example, by codes of practice. This guide from 2006 evolved into a joint FAO-OIE recommendation of 2009, which is targeted at biological, chemical and physical hazards that can enter the food-producing animal or products of animal origin. They addressed this issue in six chapters, namely (1) General farm management, (2) Animal health management, (3) Veterinary medicines and biologicals, (4) Animal feeding and watering, (5) Environment and infrastructure and, (6) Animal and product handling. As regards hazards and corresponding control points, the guide does not elaborate on specific hazardous agents, but more on events/situations likely to occur on farms, for example, under 'biohazards': 'livestock not well adapted to conditions' or under 'physical hazards': 'ingestion of dangerous/harmful objects'; 'broken needles or other penetrating objects'. For the listed hazards (or, more specifically, hazardous situations), the 'points' where the hazard occurs and can be managed, and finally, a set of recommended actions are given. For implementing the recommended actions or measures in specific production systems, it must be evaluated if the measures predefined in the guide are applicable either 'as is' or with some rewording. Such 'validated' measures are regarded as of 'critical priority' or as 'highly advisable', respectively. This implementation procedure

Table 1 Summary of the OIE Guide to Good Farming Practice (OIE, 2006), key areas and measures to be implemented. Implementation will have to consider national legislation, and some actions will require involvement veterinarians and/or official authorities

Area	Type of hazard	
Buildings and other facilities	b c p	Control of surroundings: • farming activities distant from industry likely to be a source of pollution • farm buildings separated from residential accommodation, waste storage • restrict access for visitors • farms should be separated from each other Control of environment on the farm: • livestock buildings should be adequate in size and well ventilated; surfaces must not release toxic substances • separation of clean from soiled, of storage and working areas from animal production areas • newly arrived animals and sick animals separated from the others • facilities easy to clean and disinfect • collection of waste and effluents • barriers against pests and wild or stray animals • restricted access for non-authorized persons • control measures should still work in case of occurrence of natural disasters
Health conditions for introduction of animals into the farm	b	• accept only animals on known health status (serology, vaccination etc.), and keep them separated for a period of time so as to allow acclimatization, to observe health status and carry out medical measures, if necessary (quarantine) • refuse animals showing suspicious clinical signs on delivery • animals should originate from farms working under GFP • animals must be clearly identified (tag, health document) • 'history' of animals must be documented (origin, transport etc.)
Animal feeding	b c p	Pasture and grassland • Ensure that land is suitable for pasture, esp. as regards toxic plants, residues of pesticides, pollutants from industry, parasites (e.g. liver fluke) or bacteria/bacterial spores (e.g. *Bacillus anthracis*) Commercial feed • Feed must be correctly labelled • Visual examination for signs of spoilage, esp. mould growth • Correct storage (e.g. protected from pests) • Traceability: feed batch numbers must be recorded and a feed sample be kept in case feed-related problems are suspected and tests are necessary Manufacture of animal feed on the farm • requirements listed above for 'commercial feed' apply to finished feed and raw materials, procedures and devices used for mixing and storage • composition (nutrients) should be checked regularly

(Continued)

Table 1 (Continued)

Area	Type of hazard	
		General issues
		• feed must be adequate for the animal species
		• troughs must not be overfilled and unused feed must be removed. Troughs and feeders must be cleaned regularly
Animal watering	b c	• domestic or wild/stray animals must not contaminate safe water reserves or watering points
		• livestock must not have access to polluted water
		• water reserves must be protected from pollution (e.g. by pesticides, herbicides, slurry, manure)
		• water distribution systems must be maintained and cleaned regularly
		• water quality should be tested regularly (microbiological and physico-chemical characteristics)
Veterinary drugs	c (p)	Drugs must be used on the basis of a sound diagnosis, ensuring that
		• dosage, mode of administration are correct
		• withdrawal times, if applicable, are respected
		• appropriate records are kept
Farm management		**Training, conduct and health status of staff**
		• workers must receive training to be able to do their job
		• training in biosecurity principles and practices
		• suitable working clothes and sanitary measures
		• monitoring to detect healthy carriers of biohazards transmissible to animals
		Maintenance, cleaning and disinfection of equipment, premises and immediate surroundings
		• cleaning and disinfection procedures based on manufacturer's instructions
		• tests to ascertain if procedures are effective
		Measures to control pests and stray animals and to prevent unauthorized access
		• pest control plan
		• no access of stray animals or pets to livestock buildings
		• no access for unauthorized persons
		• minimize contact of livestock with wild animals
		Stock management for feed and drugs
		Management of waste materials, effluents and expired products
		Storage of chemical products
		Production monitoring of animals
		• animals on the farm are permanently identified and records kept up-to-date
		• daily surveillance for health condition of production animals
		• production performance monitoring to detect abnormalities which could indicate the onset of disease

Health monitoring of animals and disease prevention programmes
- development and implementation of disease control programmes
- standardized operational procedures for detection, management and treatment of diseases
- take advantage of information gathered during ante- and post-mortem inspection at the abattoir
- sampling in the course of official surveillance programmes

Animal movements

Isolation of sick animals and their products
- sick animals should be separated from healthy ones
- hygiene regulations regarding contacts between humans and sick animals or animals undergoing treatment
- products from sick animals not used for food or animal feed

Storage and disposal of dead animals
- disposal in a way that pathogens cannot spread

| Preparation of animals for slaughter | b | General measures |

General measures
- animals must be fit for slaughter
- animals must be correctly identified (tagged)
- loading, unloading areas and transport devices should be kept clean to avoid soiling of animals
- animals should be handled humanely and not be stressed
- in the days prior to slaughter, the content of hay or straw should be raised
- feed be withdrawn 24h before transport
- water access should be given until transport

Extensively grazed livestock
- should be kept protected from mud at the end of the fattening phase (e.g. new pasture, watering points well distributed, or in a fenced shedded area)

Livestock housed on slatted flooring
- stocking density should be correct
- flooring be cleaned regularly
- cattle kept on straw bedding the weeks before slaughter

Livestock housed on litter
- stocking density should be correct
- clean litter must be supplied
- ventilation and drainage of effluents must be effective

Health measures
- sick animals or medicated animals in the withdrawal period must not be sent to slaughter

Common measures for record keeping and traceability

Traceability system for animals, feed and animal products
Record keeping

b, c, p refer to biological, chemical and physical hazards, respectively should allow consistent application of GFP within countries and also facilitate trade between countries. The guide should be considered in conjunction with others from the FAO on food, feed, veterinary drugs and animal handling and slaughter. In practice, a number of GFP-related practices are already part of standing legislation or are included in codes of marketing organizations, but it is above the scope of this chapter to address all national legislation and codes or guides on this matter.

There are several textbooks providing Good Hygiene Practice (GHP) elements for specific pre-harvest settings. As regards cleaning and disinfection and treatment of animal waste, the reader may refer to Motarjemi and Lelieveld (2014).

In practice, some aspects of GHP are dealt with in separate guides. One major aspect is biosecurity, which comprises preventive measures to reduce the risk of transmission of infectious diseases in agricultural products and livestock (Koblentz, 2010). In this sense, Laurence presented a comprehensive review on biosecurity measures on beef cattle farms in Australia (Laurence, 2014). The two pillars, bioexclusion and biocontainment, refer to the measures to keep pathogens and potential carriers of pathogens out and to the measures to prevent the spread of pathogens within a farm or to neighbouring farms. Similar to OIE, the author presents a list of requirements and the measures how to achieve this. For example, the requirement 'control visitors' can be achieved by the following ways: 'provide visitors with clean boots and clothes or coveralls'; 'install footbaths', 'Do not let visitors step into feed troughs or wash in water troughs'; 'limit contact by visitors with the herd'. Likewise, 'minimize vehicle traffic' can be achieved by 'keeping a separate vehicle for use on the farm' or 'prevent off-farm vehicles from driving in areas where animals travel'. Principally, the same messages are conferred to farmers in the 'National Farm Biosecurity Reference Manual – Grazing Livestock Production' (Animal Health Australia, 2012), see Table 2.

Given the differences in farming systems and farm sizes throughout the world, it is obvious that GHP at national and regional level will need some adaptions in terms of cost–benefit analyses. Perhaps the most important issue is how to communicate GFP principles to all involved parties. Expectedly, the presentation of GFP measures to the competent authority (e.g. the FAO-OIE guide from 2009) will be more comprehensive than that to specialized farmers (e.g. the grazing livestock manual, Animal Health Australia, 2012), whereas to the less knowledgeable rural population in developing countries, only the essential messages will be delivered, preferably relying on a mix of illustrations and text rather than on text only (e.g. the 'guide of best farming practices' for Haiti; OXFAM, 2014). Didactic approaches, including interaction of trainees and trainers, should be considered (see the 'Pre-Harvest Food Safety in 4-H Animal Science' training programme of the University of California, Smith et al., 2014).

Table 2 Principles to implement biosecurity for grazing beef cattle (Animal Health Australia, 2012)

Principle	What to achieve	Recommended practices (i.e. 'how to achieve') (selected)
Livestock	Manage the introduction and movement of livestock in a way that minimizes the risk of introducing or spreading infectious disease	• Check animals for health status before purchasing • Purchase animals from suppliers with a quality assurance system • Segregate and observe newly introduced animals
People, equipment and vehicles	People, equipment and vehicles entering the property are controlled to minimize the potential for property contamination	• Define permitted access areas for farm contractors • Farm contractors and visitors to use protective clothing and clean hands and boots when entering/leaving the permitted access areas • Keep a visitor register
Feed and water	Quality of stockfeed and water is fit for purpose	• Water management to reduce risks of transmitting disease agents or weeds • Ruminants have no access to restricted animal material (i.e. material from vertebrates except gelatin, tallow, milk products, oils)
Feral animals/ wildlife/weed control	Minimize the potential for wildlife and domestic or feral animals to introduce diseases to livestock	• Actions to control pests and wildlife should be coordinated with neighbouring farms • Minimize access by feral animals, domestic animals and wildlife to waste
Animal health management	Regular monitoring of livestock health to prevent control diseases	• Staff on the farm should be vaccinated against risk diseases (e.g. tetanus), • Inspection of the health of the livestock should be regular and health history of the herd be recorded
Staff instruction	All staff must understand the importance of biosecurity and implement the agreed practices	Staff knows • its role in implementation of biosecurity and knows how to recognize and handle sick and injured animals • how to contact competent authorities in case of disease outbreaks
Carcass, effluent and waste management	Disposal of dead animals and waste is managed to minimize the spread of disease	• Disposal areas not accessible for wild or domestic animals • No spread of contaminants by water

3 Animal handling and animal welfare

Perhaps the most crucial facet of GFP is that of animal welfare and animal rights. Animal welfare principles received wide interest in Western societies ca. 100 years ago, and the concept of 'animal rights' was founded only ca. 40 years ago. The basis for animal rights is that there is no justification that animals with the ability to suffer should not deserve the same consideration as humans (Singer, 1975). Basic rights for sentient animals were formulated in the 1960s, formalized by the UK Farm Animal Welfare Council in 1979 and further refined to the commonly known 'five freedoms', namely, the freedom from hunger and thirst; freedom from discomfort (e.g. by providing shelter and protection); freedom from pain, injury and disease; freedom to express normal behaviour and freedom from fear and distress. The main shift from traditional animal welfare to the five freedoms was to consider not only 'freedom from ...' but also 'freedom to ...'. These requirements are – reworded and amended – incorporated in guides of various intergovernmental bodies like the OIE (2015). The European Food Safety Authority (EFSA) has addressed animal welfare in several scientific opinions based on the five freedoms approach and using risk assessment approaches for various commodities and farm animal species (see 'further reading'). Apart from these scientific opinions, the textbook of Smulders and Algers (2009) gives a comprehensive picture of the innumerable facets of assessment of animal welfare risks and assurance of animal welfare. Since stressed or injured animals will be more susceptible to infections, or may yield meat with unfavourable characteristics (e.g. dark-firm-dry beef, DFD), it is common sense to handle animals in a way that minimizes stress and trauma. This includes, for instance, avoiding sudden and jerking movements near the head of the cattle, avoiding sudden noise and respecting the flight zone. There are a number of literature and national guides on this subject [e.g. the 'Cattle care & handling guidelines' or the 'Master cattle transporter guide' of the US Beef Quality Assurance program (BQA, 2015, undated)], but the presumably most impressive source of practical information is the website of Dr. Temple Grandin (www.grandin.com).

4 Clean cattle policy

GFP requires that cattle should be clean and not be delivered soiled, muddy and wet to slaughter. Bacterial numbers are higher on visually 'dirty' than on 'clean' hides (Hauge et al., 2015).

It could be expected that higher loads of (enteric) bacteria on cattle hides will result in higher numbers of bacteria on beef carcass surfaces, and a number of national programmes are targeted on preventing contamination of hides and grading of cattle sent to slaughter according to the degree of visual contamination. Based on this assessment, additional measures are taken (washing animals, clipping hairs, separate slaughter). In practice, assessment procedures have 4–5 categories, but there is evidence that less categories would suffice, when the relation of bacterial numbers on hides to visual contamination of hides is studied (Blagojevic et al., 2012; Hauge et al., 2015; McEvoy et al., 2000). Also, there is not always a clear relation of bacterial numbers on the skin to that on the dressed carcass, partly due to sampling procedures (in particular, which regions of the body are examined), partly due to the slaughter process as such or already implemented interventions (Hauge et al., 2015). This illustrates the general issue that not all hygiene practices in GFP need to have a direct impact on bacterial numbers. Still, it can be concluded that cattle arriving visually 'clean' at slaughter originates from holdings with a certain standard of GFP. National agencies provide guides on 'clean cattle' for farmers (e.g. FSI, 2007; SAC, 1997; TEAGASC, 2005), and illustrated grading schemes, together with suggested or required corrective steps. As an example, the main elements laid down in an UK guide are presented in Table 3. In the United Kingdom, five categories of cleanliness have been defined (Table 4). However, in practice, there will be only three ways to manage cattle at slaughter: (1) slaughter using the routine dressing procedures; (2) slaughter under extra-defined hygienic dressing controls and (3) application of extra pre-slaughter measures, for example, clipping or washing

Table 3 What to consider when implementing 'clean cattle policy' pre-harvest (SAC, 1997)

Diet	Low moisture diet, esp. in the finishing phase, correct composition of diet, no abrupt changes in diet
Housing	Slatted flooring: correct stocking density, no build-up of dung near troughs, along walls, underneath divisions Straw-bedded pens: sufficient amount of dry, clean bedding (straw) Ventilation: to ensure that bedding and cattle are dry. It is suggested to clip the backs of finishing cattle to prevent sweating
Clipping	Clipping of belly, back and tail; handling facilities required for belly clipping
Washing	No power hoses (high pressure)
Transport	No water restriction prior to transport; cattle must be dry; correct stocking density; no mixing of animals from different holdings; good ventilation; no contamination from upper to lower floors in multi-tier transporters; straw bedding instead of sawdust or wood shavings
Winter-finishing cattle	Transfer from slatted flooring to pen with bedding (2–3 weeks before slaughter); if necessary, shave before transfer; change diet gradually to dry feeds
Cattle finished off-grass	Use anthelminthics to prevent scouring from parasitic infections, respecting the withdrawal period; if young, lush grass is foraged, offer straw *ab libitum*; move feeders regularly to avoid poaching and muddiness

Table 4 Categories for assessment of cattle cleanliness (SAC, 1997; FSA, 2007)

Category	Description	Measures
1	Dry, no dung/dirt, very minor amounts of loosely adherent straw/bedding	None required
2	Dry/damp, light contamination (dung, dirt), small amounts of loosely adherent straw/bedding	None required
3	Dry/damp, significant contamination with dirt/faeces, and/or significant amounts of adherent straw/bedding	Precautions at slaughter
4	Dry/damp, heavy contamination with dirt/dung, and/or significant amounts of adherent bedding	Precautions at slaughter
5	Very wet, very heavily contaminated with dirt/dung, and/or much bedding adherent to the coat	Animal not admitted to slaughter

(e.g. SAC, 1997). As regards clipping at the slaughterhouse, clipping at lairage exerts some stress on animals (reflected in increases of ultimate muscle pH) and is a safety issue for operatives (McCleery et al., 2008).

In a nutshell, 'clean cattle' is one of the prerequisites if beef safety is to be improved by implementing interventions targeted against specific enteric biological hazards.

5 From GFP to pre-harvest food safety management: the case of *E. coli* O157

The need to augment GFP with targeted interventions became obvious with the emergence of *Escherichia coli* O157 in the beef chain in the United States (Sofos, 2002; Soon et al., 2011), and it was driven by beef processors and competent authorities as well. Some of the peculiarities of *E. coli* O157 in cattle and other ruminants are that infection will not cause signs of disease in infected animals, shedding can be intermittent, some animals will shed the bacterium in high numbers (i.e. ≥4 log cfu/g faeces; 'super-shedder'; Arthur et al., 2010) and prevalence within pens can vary considerably. This invalidates traditional methods of disease control (i.e. identification, separation and culling/treatment of infected animals; Matthews et al., 2013). The 'history' of pre-harvest management of *E. coli* O157 in cattle in the United States and the points and methods of intervention have been described by Sofos (2002). Since the problem of contamination of hides and shedding of pathogenic bacteria occurs pre-harvest, it is common sense to prevent shedding and contamination at the earliest point of the food chain instead of solely relying on post-harvest measures. The Food Safety and Inspection Services (FSIS) have issued a guidance document on this issue (FSIS, 2010), clearly defining which areas or practices are critical for control of *E. coli* O157 pre-harvest. In addition to 'principle-based husbandry' (broadly equalling GFP), interventions can be grouped into: water and feed management; water and food additives; management practices (in particular, separation of calves from adults, since calves are more likely to excrete the pathogen) and transportation; live animal treatments. As regards water and feed additives, five types of intervention were considered: (1) supplementing diet with seaweed extracts; (2) administration of ß-agonists; (3) antibiotic feed additives (e.g. ionophores), (4) probiotics (*Lactobacillus*-based) and (5), *E. coli* producing bacteriocins (colicin). Live animal treatments comprise administration of bacteriophages onto the hides/ skin of cattle, competitive exclusion by feeding non-O157 *E. coli*, washing of cattle hides with biodegradable agents, and finally, vaccination. The document was revised in 2014 (FSIS, 2014), and the array of measures was regrouped into (1) exposure reduction, (2) exclusion strategies and (3) direct anti-pathogen strategies.

Not all studied interventions are suitable in practice, and some are even conflicting, for example, fasting, which increases shedding, but reduces the amount of faeces. Vaccinations are presumably the most effective single intervention to control *E. coli* O157 on cattle farms, if at all a single intervention instead of a multiple-hurdle approach is considered. In principle, two types of vaccines are available, one targeted against the iron-transport system of Gram-negative bacteria, as *E. coli* and *Salmonella* (siderophore receptor and porins, SRP), and the others inducing immunity against type III secreted proteins of *E. coli* O157:H7. Most recently, the latter type of vaccine has been augmented by the inclusion of other antigenic bacterial structures. This may overcome the limited effect against non-O157 *E. coli*, which are most prevalent in European countries (e.g. 72% of Austrian sheep farms were contaminated predominantly with non-O157 shiga toxin-producing *E. coli* in 2013; BMG/ AGES, 2014). However, not all of these vaccines protect against colonization and/or shedding in clinical and/or field trials, or they are effective only against one serotype (independent of shiga toxin production).

A commercial vaccine against type III secretory proteins vaccines is approved in Canada, and has already been tested in other countries. It requires three injections per animal. When administered in three doses in 3-week intervals, the vaccine proved successful against experimental infections, reducing prevalence of faecal shedding by more than 50% (Potter et al., 2004). Most studies reported a significant effect in terms of reduction of prevalence of shedding (Moxley et al., 2009; Smith et al., 2009a,b), whereas Van Donkersgoed et al. (2005) found no significant effect of vaccination on in-pen-prevalence of faecal shedding. Both the different quality and the administered quantity of the vaccine may have accounted for this discrepancy. If administered, all animals of a pen need to be vaccinated to effectuate reduced shedding and hide contamination (Smith et al., 2009a). As mentioned above, a major issue is to what extent such a vaccine will be effective against different strains of *E. coli* O157 or non-O157 verotoxinogenic *E. coli* ('cross-protection'), in other words, whether the vaccine will be effective against non-O157 pathogenic *E. coli* prevalent in Europe and other regions. Nevertheless, the commercially available vaccine against type III secreted proteins has been tested in studies and has reduced shedding and colonization in the cattle of shiga toxin-producing *E. coli* O157, but has also cross-reacted with antigens of other serotypes likewise and thus has cross-protective potential against non-O157 (McNeilly et al., 2015). But further field trials are needed to prove its effectiveness.

A way to overcome this limitation is to include other cell compounds/virulence factors, as H7-flagellin or intimin. Recent research indicates that such vaccines would be very efficient at reasonable production costs (McNeilly et al., 2010, 2015).

Subcutaneous administration of an anti-SRP vaccine in two doses evoked significantly higher anti-SRP antibodies in calves, and, after experimental infection with *E. coli* O157:H7, prevalence of cattle shedding *E. coli* O157:H7 was lower than in a non-vaccinated control group. However, numbers of pathogens per g faeces were only moderately reduced (1.6 vs. 1.9 log cfu/g) (Thornton et al., 2009). Interestingly, in naturally infected cattle, a single dose was able to significantly reduce prevalence of shedding by 50% as well as the duration of shedding/high-shedding (Fox et al., 2009a,b). Efficacy of immunization was improved when calves born to vaccinated cows received a vaccination, but differences in prevalence of shedding *E. coli* O157 were not significant (Wileman et al., 2011). It must be noted that not all studies administered the vaccine thrice, as required by the manufacturer. However, a two-dose treatment combined with changes in the finishing diet was effective (Cull et al., 2012). A commercial formulation of this vaccine is conditionally approved (to be used by licensed veterinarians only).

Other types of vaccines have not been successfully introduced in the market but have proved effective at the experimental level. Most of these have not been characterized in detail for the antigenic structures they target or have targets common in many *E. coli*. Vaccines targeted against non-O157 shiga toxin-producing *E. coli* may also affect commensal *E. coli*, but these have been shown to be necessary for the development of a functioning immune system of the intestinal tract (especially the balance and the intestinal barrier between the gut microbiome and the cellular intestinal immune system) (Mirpuri et al., 2010). Vaccines that target loci directly involved in shiga toxin production or gene transfer like specific (pro)phage structures should be considered in the future.

The effect of vaccines is usually assessed by the ability to induce the formation of antibodies or by suppression of the targeted pathogens (frequency of animals excreting the pathogen and/or number of bacteria/g faeces). In case of symptomless carriage and shedding, as is the case for *E. coli* O157, the effect of vaccination is not really evident for farmers, but the additional work and costs are. From a public health viewpoint, reduction of the prevalence of shedding and super-shedding cattle by 50% is predicted to effectuate an 85% reduction in human cases of *E. coli* O157 infection (Matthews et al., 2013), but Tonsor and Schroeder (2015) could demonstrate that for farmers, this effect is not obvious or directly experienced and an economic incentive is missing to implement vaccination. In essence, this holds true for any specific measure targeted towards identification and management of super-shedders, although this would be very efficient in disease control (Munns et al., 2015).

6 Quality assurance programmes for beef production

Those involved in beef production will primarily seek advice from national organizations providing comprehensive quality assurance programmes (including certification) and training packages. The aim of such organizations is to implement best practices based on scientific knowledge and common sense husbandry techniques in cattle and beef production. To this end, information for farmers, information related to transport, etc., must be presented in a clear and concise way. In the United States, the National Cattlemen's Beef Association, linked to the 'Beef Quality Assurance' program (BQA), and the 'Beef Cattle Institute' at Kansas State University, provides such a service. Likewise, a branch of the Australian 'Livestock Production Assurance – Quality Assurance' program (LPA-QA) deals with cattle. The 'LPA Quality on-farm manual' is one example of their guides (LPA, 2014).

7 Summary

Pre-harvest food safety in beef production requires a multidisciplinary approach. HACCP-based systems can be implemented on beef cattle farms. Good hygiene practices (e.g. traceability and record keeping, bioexclusion and biocontainment, animal welfare and sustainability) form the basis that allows risk-assessment-based prioritization of hazards and the implementation of evidence-based interventions targeted against specific pathogens.

8 Future trends in research

Symptomless carriage and shedding of pathogenic bacteria is a major threat to beef safety. Despite around two decades of research on pre-harvest control of pathogenic *E. coli* in beef cattle, this issue is still not completely resolved at farm level. Although a sequence of interventions along the meat chain provide a high degree of food safety, these multiple interventions are mainly at post-harvest level and not universally portable, for example, to developing countries. Whereas toxinogenic *E. coli* in the digestive tract have been studied extensively, much less is known about other symptomlessly shed pathogens, as *Campylobacter* in the digestive tract of ruminants. It can be assumed that pre-harvest interventions will rely less on application of drugs, but more on interventions based on feed additives (e.g. probiotics), competitive exclusion and vaccination. More research on the microbiome in the digestive tract would help develop, implement and evaluate targeted control measures.

9 Where to look for further information

Further information can be taken from international bodies, as the FAO (www.fao.org) and the OIE websites (www.oie.int). Within countries, national official organizations as well as stakeholder/professional organizations provide information. Examples are: the National Cattlemen's Beef Association (http://www.beefusa.org/) for the U.S., and Livestock Production Assurance within Meat&Livestock Australia (http://www.mla.com.au/Meat-safety-and-traceability/Livestock-Production-Assurance). As regards welfare of production animals, national organizations usually provide very concise and practical training material and manuals. The website of Temple Grandin of handling of production animals is a comprehensive source of practical information (www.grandin.com). In the European Union, EFSA has issued several scientific opinions on animal welfare of production animals, and, more recently (2012), a methodological guidance to assess risks for animal welfare, by taking into account different husbandry systems, management procedures and animal welfare issues (www.efsa.europa.eu).

10 References

Animal Health Australia (2012), *National Farm Biosecurity Reference Manual. Grazing Livestock Production*. ISBN 978-1-921958-05-2.

Arthur, T. M., Brichta-Harhay, D. M., Bosilevac, J. M., Kalchayanand, N., Shackelford, S. D., Wheeler, T. L. and Koohmaraie, M. (2010), 'Super shedding of *Escherichia coli* O157:H7 by cattle and the impact on beef carcass contamination', *Meat Sci.*, 86(1), 32-7.

BQA (Beef Quality Assurance) (2015), *Cattle Care and Handling Guidelines*. http://www.bqa.org/Media/BQA/Docs/cchg2015_final.pdf (accessed 1 April 2016).

BQA (Beef Quality Assurance) (undated), *Master Cattle Transporter Guide*. http://www.bqa.org/Media/BQA/Docs/master_cattle_transporter_guide-digital.pdf (accessed 1 April 2016).

Bergschmidt, A., Nitsch, H. and Osterburg, B. (2003), *Good Farming Practice - Definitions, Implementation, Experiences*. Federal Agricultural Research Centre (FAL), Braunschweig, Germany.

Blagojevic, B., Antic, D., Ducic, M. and Buncic, S. (2012), 'Visual cleanliness scores of cattle at slaughter and microbial loads on the hides and the carcases', *Vet. Rec.*, 170, 563.

BMG/AGES (2014), *Bericht über Zoonosen und ihre Erreger in Österreich im Jahr 2013 (Report on Zoonoses in Austria, 2013)*, Federal Ministry of Health Publ., Vienna.

Cull, C. A., Paddock, Z. D., Nagaraja, T. G., Bello, N. M., Babcock, A. H. and Renter, D. G. (2012), 'Efficacy of a vaccine and a direct-fed microbial against fecal shedding of *Escherichia coli* O157:H7 in a randomized pen-level field trial of commercial feedlot cattle', *Vaccine*, 30(43), 6210-15.

FAO (2003), http://www.fao.org/docrep/meeting/006/y8704e.htm (accessed 27 January 2016).

FAO-OIE (2009), *Guide to Good Farming Practices for Animal Production Food Safety*. ISBN 978-92-5-006145-0.

Fox, J. T., Thomson, D. U., Drouillard, J. S., Thornton, A. B., Burkhardt, D. T., Emery, D. A. and Nagaraja, T. G. (2009), 'Efficacy of *Escherichia coli* O157:H7 siderophore receptor/porin proteins-based vaccine in feedlot cattle naturally shedding *E. coli* O157', *Foodborne Pathog. Dis.*, 6(7), 893-9.

FSA (Food Standards Agency) (2007), *Clean Beef Cattle for Slaughter. A Guide for Producers*, Publ. FSA, UK.

FSIS (Food Safety and Inspection Service) (2010), *Pre-Harvest Management Controls And Intervention Options For Reducing Escherichia Coli O157:H7 Shedding In Cattle*, Publ. FSIS, USA.

FSIS (Food Safety and Inspection Service) (2014), *Pre-Harvest Management Controls and Intervention Options for Reducing Shiga Toxin-Producing Escherichia coli Shedding in Cattle: An Overview of Current Research*, Publ. FSIS, USA.

Hauge, S. J., Nesbakken, T., Moen, B., Røtterud, O., Dommersnes, S., Nesteng, O., Østensvik, Ø. and Alvseike, O. (2015), 'The significance of clean and dirty animals for bacterial dynamics along the beef chain', *Int. J. Food Micro.*, 214, 70-6.

Koblentz, G. D. (2010), 'From biodefence to biosecurity: the Obama administration's strategy for countering biological threats', *International Affairs*, 88, 1.

Laurence, M. (2014), 'Biosecurity and beef cattle health, husbandry and welfare', in D. Cottle and L. Kahn (eds), *Beef Cattle Production and Trade*, CSIRO Publishing, Collingwood, Australia.

LPA (Livestock Production Assurance) (2014*), LPA Quality On-farm Manual.* http://www.mla.com.au/Meat-safety-and-traceability/Livestock-Production-Assurance/LPA-Quality-Assurance (accessed 1 April 2016).

Matthews, L., Reeve, R., Gally, D. L. J., Low, C., Woolhouse, M. E. J., McAteer, S. P., Locking, M. E., Chase-Topping, M. E., Haydon, D. T., Allison, L. J., Hanson, M. F., Gunn, G. J. and Reid, S. W. R. (2013), 'Predicting the public health benefit of vaccinating cattle against Escherichia coli O157', *Proc. Natl. Acad. Sci.*, 110(40), 16265-70.

McCleery, D. R., Stirling, J. M. E., McIvor, K. and Patterson, M. F. (2008), 'Effect of ante- and postmortem hide clipping on the microbiological quality and safety and ultimate pH value of beef carcasses in an EC-approved abattoir', *J. Appl. Microbiol.*, 104(5), 1471-9.

McEvoy, J. M., Doherty, A. M., Finnerty, M., Sheridan, J. J., McGuire, L., Blair, I. S., McDowell, D. A. and Harrington, D. (2000), 'The relationship between hide cleanliness and bacterial numbers on beef carcasses at a commercial abattoir', *Lett. Applied Microbiol.*, 30(5), 390-5.

McNeilly, T. N., Mitchell, M. C., Rosser, T., McAteer, S., Low, J. C., Smith, D. G., et al. (2010), 'Immunization of cattle with a combination of purified intimin-531, EspA and Tir significantly reduces shedding of *Escherichia coli* O157:H7 following oral challenge', *Vaccine*, 28, 1422–8.

McNeilly, T. N., Mitchell, M. C., Corbishley, A., Nath, M., Simmonds, H., McAteer, S. P., Mahajan, A., Low, J. C., David, G., Smith, E., Huntley, J. F. and Gally, D. L. (2015), 'Optimizing the protection of cattle against *Escherichia coli* O157:H7 colonization through immunization with different combinations of H7 Flagellin, Tir, Intimin-531 or EspA', *PLoS ONE,* 10(5): e0128391. doi:10.1371/journal.pone.0128391.

Mirpuri, J., Brazil, J. C., Berardinelli, A. J., Nasr, T. R., Cooper, K., Schnoor, M., Lin, P. W., Parkos, C. A. and Louis, N. A. (2010), 'Commensal Escherichia coli reduces epithelial apoptosis through IFN-alphaA-mediated induction of guanylate binding protein-1 in human and murine models of developing intestine', *J. Immunol.,* 184(12), 7186–95.

Motarjemi, J. and Lelieveld, H. (2014), *Food Safety Management: A Practical Guide for the Food Industry*, Academic Press – Elsevier.

Moxley, R. A., Smith, D. R., Luebbe, M., Erickson, G. E., Klopfenstein, T. J. and Rogan, D. (2009), '*Escherichia coli* O157:H7 vaccine dose-effect in feedlot cattle', *Foodborne Pathog. Dis.,* 6, 879–84.

Munns, K. D., Selinger, L. B., Stanford, K., Guan, L., Callaway, T. R. and McAllister, T. A. (2015), 'Perspectives on super-shedding of *Escherichia coli* O157:H7 by cattle', *Foodborne Pathog. Dis.,* 12(2), 89–103.

OIE (2015), *Terrestrial Animal Code. Chapter 7.1: Introduction to the Recommendations for Animal Welfare,* OIE, Paris.

OIE Animal Production Food Safety Working Group (2006), 'Guide to good farming practices for animal production food safety', *Rev. Sci. Tech. Off. Int. Epiz.,* 25(2), 823–36.

OXFAM (2014), *Guide of Best Farming Practices.* http://eird.org/pr14/cd/documentos/espanol/CaribeHerramientasydocumentos/Agricultura/oxfamguideagriculture2014eng.pdf (accessed 10 January 2016).

Potter, A. A., Klashinsky, S., Li, Y., Frey, E., Townsend, H., Rogan, D., Erickson, G., Hinkley, S., Klopfenstein, T. J., Moxley, R. A., Smith, D. R. and Finlay, B. B. (2004), 'Decreased shedding of *Escherichia coli* O157:H7 by cattle following vaccination with type III secreted proteins', *Vaccine*, 22, 362–9.

SAC (1997), *Technical Note T468: Producing Clean Slaughter Cattle,* SAC, Edinburgh, Scotland.

Singer, P. (1975), *Animal Liberation: A New Ethics for Our Treatment of Animals*, Harper-Collins, New York, USA.

Smith, D. R., Moxley, R. A., Klopfenstein, T. J. and Erickson, E. G. (2009a), 'A randomized longitudinal trial to test the effect of regional vaccination within a cattle feedyard on *E. coli* O157:H7 rectal colonization, fecal shedding and hide contamination', *Foodb. Pathog. Dis.,* 6(7), 885–92.

Smith, D. R., Moxley, R. A., Peterson, R. E., Klopfenstein, T. J., Erickson, G. E., Bretschneider, G., Berberov, E. M. and Clowser, S. (2009b), 'A two-dose regimen of a vaccine against type III secreted proteins reduced *Escherichia coli* O157:H7 colonization of the terminal rectum in beef cattle in commercial feedlots', *Foodb. Path. Dis.,* 6, 155–61.

Smith, M. H., Meehan, C. L., Techahun, J., Castaneda, K., Harrigan, J. and Lembrikova, J. (2014), *Pre-Harvest Food Safety in 4-H Animal Science*, University of California

Publication numbers 8508–8511. http://anrcatalog.ucanr.edu/Details. aspx?itemNo=8491 (accessed 10 January 2016).

Smulders, F. J. M. and Algers, B. (eds) (2009), *Food Safety Assurance and Veterinary Public Health. Vol.5. Welfare of Production Animals: Assessment and Management of Risks*, Wageningen Academic Publishers, Wageningen, The Netherlands.

Sofos, J. (2002), '*Approaches to pre-harvest food safety assurance*', in F. J. M. Smulders and J. D. Collins (eds), *Food Safety Assurance and Veterinary Public Health. Vol.1. Food Safety Assurance in the Pre-Harvest Phase*, Wageningen Academic Publishers, Wageningen, The Netherlands.

Soon, J. M., Chadd, S. A. and Baines, R. N. (2011), '*Escherichia coli* O157:H7 in beef cattle: on farm contamination and pre-slaughter control methods', *Anim Health Res. Rev.*, 12(2), 197–211.

TEAGASC (2005), *Producing Clean Cattle: A Guide for Farmers*. TEAGASC, Carlow, Ireland.

Thornton, A. B., Thomson, D. U., Loneragan, G. H., Fox, J. T., Burkhardt, J. D. T., Emery, D. A. and Nagaraja, T. G. (2009), 'Effects of a siderophore receptor and porin proteins-based vaccination on fecal shedding of *Escherichia coli* O157:H7 in experimentally inoculated cattle', *J. Food Prot.*, 2, 866–9.

Tonsor, G. T. and Schroeder, T. C. (2015), 'Market impacts of E. coli vaccination in U.S. Fedlot cattle', *Agricultural and Food Economics*, 3, 7. doi: 10.1186/s40100-014-0021-2.

Van Donkersgoed, J., Hancock, D., Rogan, D. and Potter, A. A. (2005), '*Escherichia coli* O157:H7 vaccine field trial in 9 feedlots in Alberta and Saskatchewan', *Can. Vet. J.*, 46, 724–8.

Wileman, B. W., Thomson, D. U., Olson, K. C., Jaeger, J. R., Pacheco, L. A., Bolte, J., Burkhardt, D. T., Emery, D. A. and Straub, D. (2011), '*Escherichia coli* O157:H7 shedding in vaccinated beef calves born to cows vaccinated prepartum with *Escherichia coli* O157:H7 SRP vaccine', *J. Food. Prot.*, 74(10), 1599–604.

Chapter 5

Biosecurity and safety for humans and animals in organic animal farming

K. Ellis, Scottish Centre for Production Animal Health and Food Safety, University of Glasgow, UK

1 Introduction

There is no generally agreed standard definition for the term 'biosecurity', although a useful description is that from the *Cambridge Dictionary* 'the methods that are used to stop a disease or infection from spreading from one person, animal or place to others'. Generally, biosecurity refers to control of infectious diseases including those which may be zoonotic; hence, reference to humans is important. Biosecurity may be further broken down into 'Bioexclusion': activities to reduce the risk of introduction, to a population, of an external infectious agent; and 'Biocontainment': activities to reduce the source and spread of an infectious agent within a population. However, it is important to note that some definitions of biosecurity include reference to biochemical substance exposure as well as infectious agents, and this broader definition has relevance to organic production systems as will be discussed later in this chapter. When considering biosecurity, the risk of introduction or exposure to an agent should be considered in the true epidemiological definition: that is risk = probability × consequence. This principle of risk can be applied to develop control measures to reduce disease or contamination of animals and foods of animal origin. All stakeholders have a responsibility to act responsibly to control disease; this includes farmers, veterinarians, animal hauliers,

http://dx.doi.org/10.19103/AS.2017.0028.05

inseminators, farm visitors, delivery drivers and so forth. The ideal is to prevent disease entering the farm, but it is inevitable that some diseases are endemic, and therefore actions must be taken, using an evidence-based approach to act to reduce the spread of disease on-farm. Detailed control measures for all diseases in a variety of species are beyond the scope of this chapter; however, there is much knowledge already in existence and the challenge is getting the knowledge to be implemented.

Biosecurity as a concept is one which has not always been embraced by farmers and, sadly, by veterinarians (Hovi, 2005). The reasons for this are complex and have been subject to a number of historical and ongoing studies, particularly focusing on the social science aspects of farming (e.g. Hovi, 2005; Brennan and Christley, 2012). The common findings with many of the studies are a wide variety in the application by farmers of different biosecurity control measures, possibly reflecting both the wide range of different control measures and perhaps stemming from a lack of a co-ordinated understanding of what biosecurity means. Other factors cited as impacting on biosecurity implementation include time and money costs, lack of proven efficacies, lack of engagement across the range of stakeholders (veterinarians, farmers, animal hauliers, etc.) and historical 'tradition' of practices such as animal markets, shows and shared/communal grazings which are high risk for disease spread. However, recent work in pig herds shows the positive impact of biosecurity implementation in terms of improved herd health and productivity and reduced use of antimicrobials in those herds with a better standard of both bioexclusion and biocontainment (Laanen et al., 2013). Currently different areas of the UK and wider European Union (EU) are concentrating efforts on control of specific diseases (e.g. Bovine Diarrhoea Virus in cattle). These initiatives include increased efforts on social science research and knowledge exchange to demonstrate the wide-ranging benefits to animals, farmers and the general public of disease control in terms of improved efficiency (reduced carbon footprint of farming), improved animal health and welfare and reductions in use of antimicrobials.

This chapter explains and explores biosecurity issues for animal farming in general, and for organic animal farming in particular, taking up the special challenges of organic animal rearing. Two case studies are used as concrete examples of how biosecurity has been addressed in practice. Many examples and both the case studies in this chapter are based on British conditions, but the principles can to a large extent be regarded as universal, although they of course have to be applied in context-relevant ways.

2 The challenges of biosecurity risk in organic farming

The ideals that form the basis of organic farming are summarised by the International Federation of Organic Agriculture Movements (IFOAM):

'Organic Agriculture is a production system that sustains the health of soils, ecosystems and people. It relies on ecological processes, biodiversity and cycles adapted to local conditions, rather than the use of inputs with adverse effects. Organic Agriculture combines tradition, innovation and science to benefit the shared environment and promote fair relationships and a good quality of life for all involved'.

These ideals are translated into regulations that are implemented differently across different international regions, often to reflect local climate or sociological factors. This means that the regional, local or even certifying body rules can differ significantly. For example in the EU, there are strict regulations on the use of veterinary medicines in organic animal farming, which include an extended 'withdrawal period' before animal products can be sent to market. However, in the USA, the use of antimicrobials is forbidden in organic animal farming. Nevertheless, there are some shared attributes in organic farming internationally that may alter the risk of disease on-farm and/or the subsequent safety of the foods of animal origin from those farms for consumers. The main areas of biosecurity risk difference in organic animal farms compared to non-organic animals are (1) the emphasis on outdoor access by animals and (2) the use of a closed biological cycle on-farm (minimising external inputs and recycling of by-products).

Despite the relatively low market share of organic produce, it is vitally important that products of organic origin fulfil the highest standards in terms of animal welfare and food safety; indeed, many consumers believe that organic foods pose fewer risks than non-organic foods and the animals are farmed to a higher standard of welfare. Many studies of consumer purchasing patterns have identified food safety, food quality and animal welfare as the main reasons for choosing organic food (Padel and Foster, 2005; Williams and Hammitt, 2001; Ellis et al., 2009). Most consumers associate organic foods with a lower risk of contamination with pesticides (Williams and Hammitt, 2001) or drug residues and reduced use of antimicrobials; however, in terms of food safety, most food-borne disease is attributed to microbiological hazards, such as bacterial or parasitic organisms, with consumers tending to underestimate the risk from common microbial pathogens such as *Salmonella* and *Campylobacter* (Williams and Hammitt, 2001; Meerburg and Borgsteede, 2010). With regard to animal welfare, organic farming, because of the emphasis on lower stocking densities and outside access, should enable farmers to provide animals with the opportunity to fulfil the need to express normal patterns of behaviour, one of the 'Five Freedoms' of animal welfare (Webster, 2001). Indeed, recent advances in animal welfare science have expanded on this point to lead to the concept of 'the life worth living' (Mellor, 2016), whereby animals not only have an existence where negative impacts are minimised, but also live in environments which provide opportunities for them to engage in behaviours they find rewarding.

Organic farming has the potential to be (and in many cases already is) at the forefront of pushing the boundaries of farming to allow animals a life worth living. However, there is a potential juxtaposition of outdoor access and the welfare benefits this may bring, versus biosecurity and food safety. Outside access makes farm-level biosecurity more complex and therefore more difficult across all farmed species, with a consequently higher risk of spread of some diseases. Prevention of disease is a vitally important component of animal welfare; indeed, it too is one of the 'Five Freedoms' (freedom from pain, injury and disease). In addition to increased disease risk to the animals themselves, it is hypothesised there could be a concurrent potential increased risk to food safety due to microbial disease. This chapter will investigate aspects of this issue, focusing on major areas of biosecurity and food safety risk.

2.1 Disease risks of outdoor production

Allowing outdoor access for organic animals is a basic tenet in the ethos of the production system and therefore is a basic requirement in the organic livestock regulations worldwide. Clearly, due to regional and climatic variations, there are differences in the duration of time spent outside and the type of outside environment available to animals. Nevertheless, outdoor access does confer the opportunity for animals to graze, forage, socialise and behave in ways very different to animals in indoor systems. This can offer the opportunity to fulfil natural behaviours and the opportunity for 'a life worth living' (Mellor, 2016), but also comes with a number of opportunities to increase risk of infectious disease, parasitic disease and access to undesirable hygiene risks, as will be discussed below. Disease can be spread via the air, direct contact between animals in neighbouring herds or co-grazed animals, contact with wildlife vectors (including airborne insects) or through water courses. Although disease-causing organisms are the same in both organic and non-organic farmed animals, outdoor access can increase disease risk as it can be harder to break infectious cycles. For example, access to grazing and an outside environment and a later weaning age can make pigs more susceptible to infection with *Cryptosporidium* and *Giardia* as pigs have more chance of contact with faeces from multiple animals (Petersen et al., 2015). Outside access inevitably leads to exposure to wildlife which can carry a variety of infectious diseases (e.g. *Salmonella*, *Campylobacter*), and organic farms often have limited or no wildlife control measures in place. Other (but not all) conditions that organic animals might be more at risk of exposure to include: *Toxoplasma gondii*, *Trichinella* spp., avian flu, leptospirosis, gastrointestinal nematodes and *Fasciola hepatica*. Some of these clearly have direct or indirect risks to food safety and/or human health, which will be discussed later in this chapter.

'Outdoor access' can be a variable definition in both organic and non-organic farming. There are varying degrees of outside access quality, which may range from an entirely outdoor existence year round, to being housed at certain times in the year to avoid extreme weather. For example, in northern Europe, dairy cows are often allowed outside access to graze in the summer and are housed in the winter. The consumer perception is very important here in driving the market towards more 'animal friendly' husbandry systems, accompanied by some very convincing marketing by the food industry, albeit that consumer understanding of animal husbandry can be very limited as much of the general public in modern society has no direct relationship with farming (Ellis et al., 2009). Depending on the species, age, breed and local climate conditions, outside access may not always be to the animal's best interest at all times (i.e. during particularly harsh weather) and it is important that this is accurately reported to the consumer, although the general public are not always open to extensive discussions on the detailed aspects of farming. Outdoor access is an emotive issue, but at all times it should be considered 'What are the needs of the animal?' and how best can we address those? One important example is avian flu, which has recently caused significant problems for outdoor poultry flocks in Europe (see Case Study 2).

2.2 The challenges of keeping animals outside

2.2.1 Parasite challenges to organic animals

Parasite exposure in farmed animals is increased when animals are allowed outside access, affecting all farmed species. In particular, animals are sometimes grazing more marginal extensive areas, which may be perfect habitats for parasites and their intermediate hosts (e.g. liver fluke in cattle and sheep). Given the emphasis on reduced use of veterinary medicines, this can be problematic in some farms, due to combined circumstances of local micro-environment of the farm, grazing management and inappropriate strategic use of available flukicides and/or anthelmintics (Ellis et al., 2011; Jackson, 2012). Recent developments in the use of more targeted anthelmintic therapy, largely to address increasing anthelmintic resistance, show potential benefits in particular for organic producers. In particular, the Targeted Selective Treatment approach has been shown to be applicable on commercial farms (Busin et al., 2014) and has been used in both organic dairy and beef herds (Jackson, 2012). Although these approaches require increased handling of the animals to enable individual animal responses to a variable parasite challenge to be assessed, the resultant targeted (and therefore usually reduced) use of anthelmintic is entirely in keeping with the organic ethos.

One of the major zoonotic parasitic diseases worldwide, in terms of distribution and potential effects, is *Toxoplasma gondii* (Meerburg and Borgsteede, 2010; EFSA, 2016). This parasite can infect a number of farmed species including sheep and pigs, although from a food safety perspective, contaminated pork is one of the main sources of human infection, particularly due to the relatively high consumption of pork meat (Slany et al., 2016). Increased infection rates are reported in outdoor reared and organic pigs, with the prevalence of infection in pigs increasing with increased time outside (Slany et al., 2016; Wallender et al., 2016). Suggestions have been made to control the disease by reducing exposure to cats (and cat faeces) and to vermin such as wild rodents, although in practical terms, this can be very difficult and it may be difficult for farms allowing pigs outdoor access to become 'low risk' for *Toxoplasma* infection (Wallender et al., 2016). Instead, given the wide distribution of the parasite in different farmed species, from both organic and non-organic farms, it may be that freezing meat from outdoor herds and ensuring the important food safety message to consumers is the proper cooking of meat before consumption are the two most pragmatic action points.

Recent work on *Cryptosporidium* and *Giardia* infections in organic pigs in Denmark found variations in infection intensity and prevalence between different age groups of pigs in the study herds, all of which were higher than reported rates in non-organic pigs (Petersen et al., 2015). The authors report that this is inevitable given the outdoor rearing, but the risk to human health is probably minimal as the species of *Cryptosporidium* and *Giardia* that were isolated were in the majority pig-specific and not zoonotic. Differences in management between organic and non-organic dairy herds, such as leaving calves with cows for longer, were hypothesised to affect infection rates with *Cryptosporidium* in dairy cows and calves in Sweden; however, a similar prevalence between organic and non-organic herds was detected in a recent study, thus presenting a similar zoonotic risk (Silverlås and Blanco-Penedo, 2013). Further analysis of data from that study suggested that factors associated with infection included bedding cleanliness, pen cleaning and biosecurity awareness of the farmers.

Many parasitic disease risks can be mitigated by application of biosecurity protocols on farms and particularly bioexclusion: the avoidance of buying-in animals (which may be bringing new parasites or anthelmintic resistant parasites) and biocontainment: the use of evidence-based approaches to restrict endemic parasite spread. Nevertheless, some diseases are very difficult to avoid if there is a wildlife reservoir and outdoor animals can contact infected wildlife, or wildlife faeces, in which case a HACCP (Hazard Analysis and Critical Control Point) approach should be adopted to protect food chain safety.

2.2.2 Organic eggs and residues

Human exposure (including from food) to dioxins has been shown to have a range of effects depending on exposure dose, from many acute disease presentations through to increased cancer risk. Research among layer flocks in various EU countries has shown that organic eggs can contain higher concentrations of dioxins than non-organic eggs (De Vries et al., 2006; Piskorska-Pliszczynska et al., 2015). The hens' intake of dioxins can be from various sources acquired outside, including soil, plants, access to rubbish and invertebrates. Due to the free-ranging nature of organic hens, they inevitably make more use of these food sources than non-organic birds. The main factor thought to increase dioxin intake risk is consumption of invertebrates and soil. Management interventions to reduce exposure to these risk factors are difficult as they inevitably compromise the aims of the organic systems. Additional work in this field (Luzardo et al., 2013) suggests that the persistent organic pollutant contamination of eggs from different production systems can be highly variable between and within system and it is likely that individual farm-specific factors have highly variable impacts on the eggs.

2.2.3 Zoonotic bacterial infection risks

Salmonella and *Campylobacter* are two zoonotic bacterial pathogens responsible for the majority of reported food-borne illness in both the USA and the EU (Salaheen et al., 2015; EFSA, 2016). Poultry and poultry products are the main source of infection, usually acquired through improper handling of raw product or incomplete cooking (Salaheen et al., 2015). *Salmonella* can infect a variety of farmed (and wild) animal species and may be difficult to detect as it can often be asymptomatic in the host animal. Various studies have looked at *Salmonella* infection rates in organic and non-organic farmed animals with varied results: higher, no difference and lower infection rates have been described in organic farmed broilers and pigs compared to non-organic broilers and pigs (Alali et al., 2010; Meerburg and Borgsteede, 2010; Salaheen et al., 2015; Helke et al., 2016; Peng et al., 2016). At environmental level on-farm, retail and slaughterhouse level, organic poultry samples have mostly been reported to be more likely to be positive for *Salmonella* (Hardy et al., 2013; Salaheen et al., 2015; Peng et al., 2016) although there has been some difference reported in *Salmonella* species diversity between farms (Peng et al., 2016). In eggs laid by hens from different systems of housing laying hens, organic eggs have been reported to have higher microbial loads compared to non-organic eggs (Galis et al., 2012), with a concurrently higher isolation prevalence of *Salmonella*.

Campylobacter infection rates have been found to be higher in organic broilers and in organic pigs, with a shift occurring with outdoor exposure to increase the prevalence of *Campylobacter jejuni* found in pig faecal samples (Meerburg and Borgsteed, 2010; Hoorebeke et al., 2012). The ecological dynamics of *Campylobacter* in mixed crop livestock (MCL) farms has been further investigated, with a higher prevalence of positive samples found from a range of faecal, environmental (compost, soil, grass, bedding, invertebrates) and feed samples from MCL farms compared to conventional farms (Salaheen et al., 2016). Aside from outdoor access and potential infection from wild animal reservoirs or water sources, infection with *Campylobacter* may be more likely in poultry production due to longer rearing periods (allowing more time for colonisation), variation in vulnerability of different slower growing strains or types of bird and exposure to, and eating of, invertebrate populations which can harbour *Campylobacter* (Salaheen et al., 2015; Ahmed et al., 2016). Again though, there is some contradictory evidence, where some studies report no difference in *Campylobacter* infection prevalence in different rearing systems of birds (Hanning et al., 2010). Some recent work does suggest that use of fly control (fly screens) in extensive poultry systems can reduce *Campylobacter* carriage (Bahrndorff et al., 2013) and these relatively simple, albeit labour requiring steps, can be useful in organic farms. At retail level (both farmers' markets and supermarkets), MCL and organic poultry samples have shown a higher prevalence of contamination compared to non-organic, albeit that poultry from all source farm types had an increased prevalence of contamination, reflecting the cross-contamination that occurs in all poultry slaughter and processing units (Salaheen et al., 2015, 2016).

2.3 *Potential disease risk in a closed biological cycle*

In addition to the previously discussed point that organic reared animals have the opportunity to go outdoors and may be more likely to contact infectious disease, parasites or environmental toxins that may be of significance to human health, organic farming also places great emphasis on working with biological cycles and reusing nutrients on-farm. In practice, this means extensive use of animal manures to fertilise crops and pasture; albeit this also occurs on non-organic farms, non-organic producers have a variety of synthetic fertilisers available to use; so the relative emphasis of manure use is greater on organic farms. Organic farms often adopt an MCL farming approach, which may be spatially separated, rotational or fully integrated (Salaheen et al., 2015). An efficient MCL system can give considerable benefits to the soil structure and fertility; in efficiency of resource use on-farm and animals can act as natural pest control (e.g. poultry can eat invertebrates). There is considerable evidence showing that soil treatment with raw manures/incompletely composted manure that harbour enteric zoonotic pathogens (i.e. *Salmonella, Escherichia coli* 0157) is a potential means of introducing these pathogens into the food production system (Meerburg and Borgsteede, 2010). Additionally, crops can be contaminated from surface water run-off, irrigation and indirect spread from animals via vectors (Salaheen et al., 2015). Less is known about the actual reality of this occurring on-farm. In a review of recent evidence, Meerburg and Borgsteede (2010) hypothesised that organic farming may present a higher risk of contamination of crops and feedstuffs with zoonotic pathogens; however, the experimental data are varied and contradictory. Although animal manure could introduce pathogen contamination to foods, it should not be considered a common problem, but that the presence of pathogens should always be considered and appropriate steps taken to reduce those risks (Leifert et al., 2008). Some composting methods can reduce significantly, or destroy pathogenic microorganisms in animal manures; however, in practice, composting does not always occur to the degree required (reaching high enough temperature) to destroy pathogens, particularly when climate variables, such as ultraviolet light exposure (or lack of) and moisture presence or absence (Salaheen et al., 2015), are included.

2.4 Other biosecurity risks

2.4.1 Animal transport

A major disease risk for all farmed animals is when they mix with animals from different farms. This can occur during markets, shows, in rearing units or in communal grazings and, arguably, such risky activities should be reduced. However, animal shows and markets are important in a social context for many farmers and they offer a window on the individual farm and the wider industry, which is not only of benefit to those farmers wishing to sell animals, but can be an important opportunity to showcase well-cared-for animals to the public and therefore has significant educational potential. Organic regulations place emphasis on good welfare standards when transporting animals, with organic farms being able to move animals to market and to rearing/finishing farms similar to non-organic farms. A key point here is for organic producers to discuss with their veterinarian the potential risks of shows, sales and buying-in animals and should have in place an appropriate, farm-specific set of biosecurity protocols, covering aspects including but not limited to quarantine of animals, sourcing of animals, testing of animals and so forth, to reduce the risk of disease transmission.

2.4.2 Emerging diseases and changes in disease patterns

New and evolving patterns of disease that affect animals are occurring worldwide, for example Schmallenberg disease and the spread of Bluetongue, respectively. These diseases are vector-borne and animals kept both inside and outside have been affected, albeit the risk is much higher for animals kept outdoors (MacLachlan and Mayo, 2013). Realistically, it is very difficult for farmers to protect their animals completely from biting insects and here, the use of vaccination (currently available against Bluetongue) is a key control method.

Climate change is widely accepted by many scientists as a real phenomenon, and certainly new patterns of rainfall and temperature have been recorded in many parts of the world. This can have an effect on disease epidemiology, in particular those parasitic diseases where part of the parasite lifecycle occurs outside the host (with or without an intermediate host). This has been demonstrated by changes in the patterns of disease recorded, for example for liver fluke (*Fasciola hepatica*) associated with warmer, wetter weather in the UK (Fox et al., 2011). A wider distribution of parasite spread and longer periods of risk at pasture can contribute to increased disease in outdoor animals. The key to understanding and controlling these diseases, both new/ emerging diseases and geographic spread of existing/known diseases, is to ensure there is appropriate disease surveillance, so that diseases can be diagnosed and action can be taken to reduce the risk of further disease by the most appropriate means or combination of means (e.g. vaccination, strategic drug use, pasture management).

2.4.3 Open farm access

One aspect of the IFOAM organic principles of social inclusion that may potentially pose a risk to public health is the concept of open farms and encouraging the public/volunteers to access working animal farms. Undoubtedly, there is a massive educational potential in being able to show the public where their food comes from and getting engagement in a range of aspects of food production, animal welfare and sustainability; however, it must be acknowledged that there is a risk of zoonotic disease on even the most fastidious farm and these risks must be appropriately conveyed to the visiting workers and/or public.

2.5 Antimicrobial resistance

Organic regulations worldwide seek to limit the use of antimicrobials in animal production and in some countries such as the USA, their use is banned in organic farming. This leads to the hypothesis that the prevalence of antimicrobial resistance (AMR) in bacteria from organic farms should be lower due to reduced selection pressure. There are considerable data on this subject area: one of intense public and political scrutiny currently, and the general trend would suggest that there is a lower prevalence of resistant bacterial species in organic farms. However, there are a number of unexpected or anomalous findings which cannot be ignored, highlighting the complexity of this area of research. Multidrug resistant bacteria have been found in organic farms, contrary to expectation. Although the main determinant of antimicrobial susceptibility is exposure to that particular antimicrobial, there are other factors/pressures which may lead to the expression of resistance genes in a bacterial population and this is an area of continued research development which will be discussed further in this chapter. A universal policy of reducing indiscriminate antimicrobial use across all farm types is one which should be supported. This requires a measured, evidence-based approach to antimicrobial use on all farm types (for example see Allen and Bellini, 2017).

A recent US work examining the antimicrobial susceptibility of *Campylobacter* isolates obtained pre- and post-harvest from MCL, (including organic in this definition) and conventional farm sources found isolates from the MCL being more susceptible to a range of antibiotics except, interestingly tetracyclines (Salaheen et al., 2016). This may be due to cross-contamination with pig faeces occurring on MCL farms, but is an area worthy of further investigation. Tetracycline resistance has been studied in microorganisms isolated from organic and non-organic beef, pork and poultry (Guarddon et al., 2014), with data suggesting the wide distribution of some tetracycline resistance genes, independent of the degree of use of antimicrobials on-farm.

Various studies of poultry and poultry products have found a reduced prevalence of antimicrobial resistant bacteria including *Escherichia coli* and *Salmonella* isolated from organic samples (Alali et al., 2010; Sapkota et al., 2011, 2014; Alvarez-Fernandez et al., 2013; Peng et al., 2016). Studies of laying hens have reported lower AMR rates in *Enterococci* (used as an indicator bacterium) and a number of gram-negative bacteria including *Salmonella* (Schwaiger et al., 2010; Alvarez-Fernandez et al., 2012). In layer rearing units, *Campylobacter* infection rates have been found to be similar between organic and non-organic farms, but the AMR prevalence was lower in organic flocks (Kassem et al., 2017).

Despite a reduced usage of antimicrobials on organic broiler and pig units, antimicrobial resistant bacteria of different species have been isolated. A study from Holland found a similar prevalence of extended-spectrum and AmpC β-lactamase (ESBL) producing *Escherichia coli* in both live birds and people living and/or working on the farms (Huijbers et al., 2015). Multiresistant *Clostridium perfringens* isolates have also been found in organic broiler faecal samples (Alimolaei et al., 2015). Fluoroquinolone-resistant *Campylobacter* have been isolated from organic poultry at slaughterhouse level (Fraqueza et al., 2014). A multinational study (Sweden, Denmark, France and Italy) of organic and non-organic pig farms (Gerzova et al., 2015) found no difference in AMR genes in microbiota originating from the different farm types; indeed, there was more variability between geographic location of the farm rather than farm type. In contrast, an additional study of organic pig farming in the same four countries (Osterberg et al., 2016) found a lower AMR prevalence in bacterial samples from organic herds, although again, this highlighted significant differences in AMR prevalence between different countries. These findings were unexpected by the authors and highlight the deficiencies in making dogmatic statements about any farming practice.

The AMR profile of cattle farms has been investigated to a lesser extent, although evidence suggests that organic farmed cattle are associated with bacterial isolates with reduced AMR compared to non-organic cattle (Berge et al., 2010). Again though, there are other factors at play which appear to have some bearing on AMR prevalence including age of cattle sampled (more likely in calves rather than adults) and geographic location of the farm. Interestingly, one study from the USA found a reduced prevalence of AMR genes and bacteria from environmental samples that captured airborne bacteria near organic and non-organic beef cattle farms in California (Sancheza et al., 2016). These data are very interesting in terms of potential significance to public health and spread of AMR potential.

Studies of dairy herds have found variable bacterial populations in milk samples, including isolation of methicillin-resistant *Staphylococcus aureus* from organic cow milk (Cicconi-Hogan et al., 2014). There is a suggestion that reducing the antimicrobial use can have relatively quick effects on AMR in mastitis pathogens, as decreasing resistance was recorded in samples taken over periods ranging from six months to three years from dairy herds transitioning to organic farming (Suriyasathaporn, 2010; Park et al., 2012). Milk samples from dairy goats and sheep in Greece have reported a lower AMR rate compared to non-organic samples (Malissiova et al., 2017). In contrast, detection of extended-spectrum beta-lactamases (ESBL)/AmpC *Escherichia coli* in faeces from a proportion of Dutch organic dairies (Santman-Berends et al., 2017) did not show a direct relationship between degree of antimicrobial use on farm and ESBL detection, suggesting that other factors, as yet unqualified, are likely to be having effects.

Overall, there is a trend for samples from organic farms, direct from the animals, from the environment or from end product samples, to show a reduced prevalence and degree of AMR. However, there is evidence that AMR can still be detected in bacteria from organic farms, and it is important that the intricacies of this subject area are fully appreciated before dogmatic statements are made. Recent work has suggested that even the application of manure (from cows that had not received antibiotics) can affect the abundance of AMR bacteria compared to soil treated with inorganic fertiliser (Udikovic-Kolic et al., 2014). Thus, there is much we still have to learn. Nevertheless, given the pressure on agriculture to reduce the use of antimicrobials, conventional farms may well be able to learn from well-managed organic farms.

3 Food safety summary

Given the potential, but not consistent increased risk of contamination of organic foods with some zoonotic parasites and bacteria and the concurrent, but not consistent trend for organic farming to be associated with reduced AMR prevalence, the consumer has to decide what he/she thinks is the most important: animal welfare opportunities and reduced AMR vs. potential increased food hazard risk. However, given the dissociation that most consumers have from agriculture, the general public are not renowned for their rational decision-making with respect to food sourcing. Currently, there are gaps in knowledge about the benefits or risks of organic foods to human health (Williams and Hammitt, 2001). Analysis of data from human disease outbreaks is limited and often the exact source of a zoonotic disease cannot be attributed (EFSA, 2016). The overwhelming current consumer perception is that organic foods are healthier in particular with respect to pesticides and residues, and 50% perceive that organic foods also have a reduced risk due to expected lower contamination with toxins and microbial pathogens (Williams and Hammitt, 2001).

To determine whether produce of organic origin was a significant public health problem, data from food-borne disease outbreaks would need to be reviewed. This would tend to capture those diseases of microbial origin rather than long-term exposure issues, which are much harder to quantify. Few surveys of farmer's market produce have been conducted, although some disease outbreaks have been back traced to farmer's market origin (Salaheen et al., 2015). In MCL farming and/or organic farming, post-harvest processing of animal products may be done at a relatively small scale in more basic facilities, using a variety of techniques and disinfectant chemicals to reduce cross-contamination. As a result, the prevalence of food-borne pathogen contamination may be highly variable (Salaheen et al., 2015). The food safety/trading standards legislation pertaining to such markets varies from country to country; for example in the USA, small farm operations retailing produce in a limited geographic area have been exempt from some aspects of food safety legislation (Salaheen et al., 2015). In combination with the lack of determination of 'source' in food-borne disease outbreaks in many cases, there is, therefore, a gap in the knowledge base to inform stakeholders on the risks associated with organic foods to the consumer.

4 Controlling infectious diseases

There are a number of strategies that can be employed by producers and processors to reduce infectious disease spread both on farm and in the food processing chain. On farm (pre-harvest), a thorough understanding of the epidemiology of the diseases of concern is vital before control measures can be recommended. It should be remembered that each farm is unique, with its own set of circumstances which should all be considered; these include: geographic location, climate and microclimate, adjacent farms, animal species and breeds held on farm, animal management, stockmanship, farm production goals and diversification enterprises on farm. Ideally, in controlling diseases of animals, there should be a team approach including the producer (and all staff on farm) and the farm's veterinarian, in combination with additional experts dependent on farm enterprise type, which may include nutritionists, grassland management advisors, semen suppliers and so forth. The end market for the finished product should also be kept in mind: supermarket standards requirements, market requirements, consumer feedback and requirements if selling direct, and so forth. This approach would be defined as herd or flock health in the true sense, and requires a constant reappraisal (the Herd Health Cycle) to record accurate production or disease data, make appropriate interventions and monitor response. This is a constant challenge, to maintain interest and enthusiasm on all parties, but can be extremely rewarding.

For specific diseases, vaccinations are available with proven efficacy to reduce disease prevalence on farm, for example *Salmonella* vaccines, which are licensed for cattle and poultry, and Bluetongue vaccines for ruminants. Vaccines against parasitic diseases are less numerous as effective vaccines against parasites are extremely difficult to produce. Two notable exceptions are the lungworm vaccine for cattle and the relative recently licensed *Haemonchus* vaccine for sheep. There is variation worldwide in the availability of licensed vaccines for animals, and this should be considered at the local level when planning herd or flock health protocols. There are some within the organic sector who do not support the free use of vaccination as they have concerns regarding the safety or ethical suitability of vaccination, in particular with respect to the adjuvants or co-products that may be included in vaccines. There are no data currently available to make an evidence-based judgement on these concerns with respect to animal vaccines. An ethical discussion, by definition, often has less clear evidence; it is a person's opinion. This is an area that is open for further discussion and many people have very strong views on this, which are also sometimes overridden by national or international legislation about compulsory vaccination schemes. Currently, vaccines are allowed by many organic certification bodies 'where there is a known disease risk', although this is not always easy to define. There is a particular concern if farmers cease vaccination, thus leaving naïve animals and there is no change in either management practice to reduce disease risk, or change in national disease status, thus leaving a potential welfare disaster to unfold.

There is a field of work investigating the microbial 'microbiome': the diversity of the microbial ecosystem in specific functional body regions (e.g.

the intestine and the udder) across a range of species (Salaheen et al., 2015). Some work suggests that probiotics (beneficial microorganisms), prebiotics (non-digestible foodstuffs that stimulate beneficial microbe growth) and phytobiotics (plant-derived chemicals), when included in feeds, can improve health and productivity and reduce carriage of pathogenic microbes (Salaheen et al., 2015). This is an area of rapid experimental development concurrent with laboratory and data analytical developments that facilitate the handling of large data sets. This may lead to future recommendations to improve animal health in the absence of use of antimicrobials or other veterinary products and may be highly applicable to the organic sector.

At a higher level, disease control needs a genuinely holistic teamwork approach throughout the agricultural industry. There needs to be engagement of stakeholders across the agricultural sector with veterinary expertise including both local practitioners and veterinarians advising industry/government, animal scientists and most importantly, industry champions (farmers, hauliers, markets, shows, retailers) that demonstrate best-practice or innovative solutions to farming challenges. Often, disease control requires some degree of legislative backing in order to support the aims. One example of this is the Scottish Bovine Viral Diarrhoea eradication scheme (see Case Study 1).

5 Conclusions and future trends

Animal disease and food safety risks are inherent in all farming systems, and producers should strive to reduce those risks wherever and whenever possible, both to improve the quality of life of the animals and to reduce the risk of infection with zoonotic pathogens. Organic systems present certain challenges associated with the emphasis on outdoor access for animals and the increased emphasis on the use of animal manures as fertilisers. As ever, there are no 'silver bullets' that reduce disease risks and the emphasis should be on a truly holistic approach that strongly values the benefits of excellent stockmanship and a progressive approach to the use of veterinary herd and flock health planning, appropriate disease control measures, including evidence-based strategic therapeutic treatments, risk-based use of vaccination, disease eradication and herd and flock health monitoring. Organic farming is associated with a reduced prevalence of AMR, most likely associated with the reduced selection pressure due to reduced antimicrobial use on-farm. There are many, well-run organic farms that can demonstrate high standards of animal health and welfare and can be held as example farms to the wider farming industry. The organic farming sector is highly valued by consumers in its commitment to high standards of animal health, welfare and food quality; it cannot allow for compromise in these areas, either for its own sake as this would be contrary to the IFOAM ideals, or in terms of long-term sustainability as a viable agricultural prospect.

There is increasing work worldwide on making antimicrobial and anthelmintic use more discriminate in order to slow down the rate of resistance development from bacteria and parasites. Organic farming can lead the way in this field as many of the tools and techniques to manage animals in the absence (or greatly reduced use) of these drugs are already there. There must be a joined-up approach to animal health that includes new developments in microbiome technology and understanding, social sciences to look at motivations for behaviours (and behavioural change) and education of consumers and the wider agricultural industry to understand the effects of certain practices. Diseases and pathogens will continue to evolve and emerge, and the animal farming sector must evolve in parallel.

6 Case studies

6.1 Case Study 1: BVD virus control in Scotland

The Scottish cattle industry, supported by Scottish Government, embarked upon a Bovine Viral Diarrhoea (BVD) eradication programme in 2010. The impact of this programme has been significant: the majority (90%) of Scottish breeding holdings now have a negative BVD status. In addition, many farmers now report reduced animal losses, particularly in suckler beef herds, due to the removal of the immunosuppressive effects of the BVD virus. Reducing BVD prevalence is likely to reduce the use of antimicrobials on-farm as the national herd has a better overall immune function. Reducing the losses at farm level will also reduce the carbon footprint of cattle farming in Scotland as farms are becoming more efficient.

BVD is an economically important disease of cattle worldwide and causes a wide range of cattle health problems such as abortion, infertility, respiratory and gastrointestinal disorders. In 2014, it was estimated that eradication of BVD in Scottish beef suckler farms would generate an average saving of around £2000 per typical herd per annum through avoidable loss in output. The success of the BVD eradication programme in Scotland is perhaps because it is an industry-driven initiative which is supported by Scottish Government legislation [The Bovine Viral Diarrhoea (Scotland) Order 2013]. The programme comprises five phases:

Phase 1) subsidised screening (2010–11),
Phase 2) mandatory annual screening (2013),
Phase 3) control measures (reducing the spread of infection) (2014),
Phase 4a) enhanced testing and further movement restrictions (2015),
Phase 4b) positive BVD status for herds with a living persistently infected (PI) (10 April 2017 – ongoing),
Phase 5) Scottish Government public consultation, Summer 2017.

Overall, the programme has made a significant impact: the majority (90%) of Scottish breeding holdings now have a negative BVD status.

Disease eradication programmes are notoriously difficult when it comes to the final phase of eradication. As the number of infected animals and infected herds decreases, it gets more difficult to find and remove them; persuading farmers to modify their behaviour in order to achieve disease freedom is critical. BVD eradication is particularly challenging due to the existence of PI cattle which appear healthy, but continue to shed virus and cause infection in other animals. PI animals may show no clinical signs as calves, although they may have a reduced growth rate and productivity and their life expectancy is significantly reduced. In January 2017, there were 410 known PIs remaining on 167 holdings in Scotland.

As the prevalence of BVD decreases, the Scottish national herd is increasingly susceptible to reinfection. Therefore, the careful use of vaccination in combination with biosecurity on-farm is vital to maintain healthy cattle. Social science is an important aspect of disease control in animals. Understanding the role of farmer behaviour is vital to understand farmers' on-farm activities and decisions about biosecurity and compliance with BVD policy. On-farm biosecurity that is practical and applicable, taking into account aspects such as animal shows, is vital. The importance of biosecurity in controlling the transmission of BVD is well recognised. Introducing BVD into a herd through fomites, contact with neighbouring stock and bringing stock onto the farm are risks that good biosecurity practice can help mitigate. In conjunction with semi-structured keeper interviews, social science research conducted by the EPIC group (Epidemiology, Population health and Infectious disease Control), on farmer behaviour, is helping to offer insights into better ways to help farmers reduce the risk of bringing BVD onto their farm.

References:

Voas, S. (2017). Scotland's BVD eradication scheme: An update. *Veterinary Record*, 180, 451–2.
EPIC website: http://www.epicscotland.org/
This website brings together advice on disease epidemiology and control, including sections for vets, farmers and policymakers.

6.2 Case Study 2: housed poultry for avian flu control

In 2016 and continuing in to 2017, there were a number of outbreaks of the highly pathogenic H5N8 avian flu strain in commercial poultry farms in Europe. In the UK, control measures included a period of legally enforced housing of poultry which aimed to reduce the risk of transmission from wild birds. This was particularly problematic for many small poultry flock keepers, including backyard hens and commercial organic farms. Although for a temporary period, housed hens were still allowed to have their eggs sold as 'free range', this period was limited under EU law to a maximum period of 12 weeks and any hens remaining housed beyond the period when restrictions were partially lifted at the end of February 2017, were not able to be called 'free range'.

Even when enforced housing restrictions were removed, keepers were subject to legal requirements to comply with restriction (prevention) zone areas. These restriction zone areas were determined by likely risk of avian-flu-infecting hens. For many keepers, choosing to keep their hens indoors continued to be the best way to comply with the requirements of the restriction zones and protect their hens from disease. Free-range producers had to be aware that products from housed hens may no longer qualify as free range.

Other than housing or keeping hens in netted enclosures, all poultry keepers, even those outside protection zones, were recommended to take 'appropriate and effective biosecurity measures' to protect hens against contact with wild birds. The Scottish Government produced a checklist of steps keepers should consider, although in reality some of these steps are not all that practical/ pragmatic. Such measures to improve biosecurity include those shown in Box 1.

Box 1 Measures to improve biosecurity

Before allowing poultry or other captive birds to use a range after a period of housing, the range must be checked and any obvious contamination from wild birds (such as faeces or feathers) must be removed.

Discourage wild birds from using range areas, e.g. through the use of (wild) bird-scarers, decoy predators and/or netting smaller range areas.

Inspect your range regularly and remove any obvious contaminants from wild birds (such as faeces or feathers) in a biosecure manner.

Operate effective barrier hygiene before entering a house or biosecure area on the premises (e.g. coveralls and dedicated boots that are only used in particular houses or bird areas).

Routinely clean and disinfect any concrete walkways on-site.

Additionally keepers were encouraged to try to make their hens' range area unattractive to wild birds. This included:

Netting ponds and draining waterlogged areas of land. If this is not possible, then fencing them off from hens so that they cannot access it whilst ranging, or using an alternative paddock that does not have access to water.

Removing any feeders and water stations from the range, or ensuring that they are covered to sufficiently restrict access by wild birds.

Considering using decoy predators or other animals (such as sheep or cattle) on the range, or allowing dogs around the range.

These aspects of biosecurity are clearly very difficult in practice to implement, and keepers of poultry must abide by legislation to protect wildlife from unnecessary disturbance. From an organic perspective, it is counter to the ethos of supporting biodiversity on-farm, to then be actively discouraging wild birds from the farm environment. Clearly, this places keepers in a very difficult situation morally: to protect the health and well-being of the animals under their care, but to be actively trying to discourage wildlife. There are no easy answers. Poultry keepers were directed to appropriate national wildlife preservation bodies to help give advice, in particular if the farm was in, or close to an area of special wildlife interest.

Organic producers were advised that the organic status of poultry flocks is not affected by any legal requirement to house or restrict access to open-air runs, provided that all other requirements of the specific certification body's organic standards continued to be met. In order to maintain animal welfare in housed flocks, advice was given on how to provide resources for housed hens to enrich their environment. Examples given were to hang objects like bird toys, twigs and cabbage or kale leaves from perches or the ceiling of the enclosure for them to peck, or provide foraging items inside like hay, dirt clumps, (non-toxic) weeds or old wood stumps, ensuring that these had not come into contact with wild birds or their faeces. In this process to protect domestic poultry from disease, it was (and remains) vitally important that the welfare of the hens is not compromised in any way, particularly as public perceptions of forced housing can be very mixed. Further information on animal welfare in housed hens can be found in the UK Government guide: *Biosecurity and preventing welfare impacts in poultry and captive birds*. Available at: www.gov.uk/government/uploads/system/uploads/attachment_data/file/616222/captive-birds-biosecurity.pdf.

Additional more general biosecurity advice was also updated for hen keepers (for example see: http://www.gov.scot/Resource/0049/00492296.pdf). However, avian flu remains a significant challenge to the poultry industry and is likely to be a recurrent one in the future. Poultry keepers may have to learn to live with the constant threat of a disease which seems to be more prevalent in the wildlife reservoir and when it is so difficult to protect free-ranging birds from contact with wildlife. The welfare benefits to poultry are enormous when they are allowed to live in well-managed outdoor flocks and it would be disappointing if this was compromised by disease control on a regular basis, albeit from well-meaning intentions to protect them from disease. This is a challenge the organic poultry industry will need to manage.

7 Where to look for further information

7.1 IFOAM website: www.ifoam.bio

Worldwide organisation linking every aspect of organic food and farming.

7.2 Farm Biosecurity, Australia

This Australian website gives useful basic facts around risks that may affect animal health and preventive measures to protect animal health. There are a number of informative short videos to illustrate practical biosecurity measures across a range of farm types (large and small): http://www.farmbiosecurity.com.au/.

7.3 Defra Farm Biosecurity, UK

This is the UK Defra (Government) website giving information on disease control across a range of farmed species (https://www.gov.uk/guidance/controlling-disease-in-farm-animals).

7.4 Moredun website: http://www.moredun.org.uk/

The Moredun Research Institute is working on a number of ways to sustainably control gastrointestinal parasites in animals, as well as a number of other animal diseases.

7.5 Bristol AMR link: http://www.bristol.ac.uk/vetscience/research/infection-immunity/main/

AMR research at the School of Veterinary Sciences is promoted and facilitated by the AMR Force. Working nationally and internationally, the project is interested in decreasing antibiotic use while improving animal health through a plurality of approaches addressing differing styles and attitudes.

8 References

Ahmed, M. F. E. M., El-Adawy, H., Hotzel, H., Tomaso, H., Neubauer, H., Kemper, N., Hartung, J. and Hafez, H. M. 2016. Prevalence, genotyping and risk factors of *thermophilic Campylobacter* spreading in organic turkey farms in Germany. *Gut Pathogens*, 8, 28.

Alali, W. Q., Thakur, S., Berghaus, R. D., Martin, M. P. and Gebreyes, W. A. 2010. Prevalence and distribution of *Salmonella* in organic and conventional broiler poultry farms. *Foodborne Pathogens and Disease*, 7, 1363–71.

Alimolaei, M., Ezatkhah, M., Bafti, M. S. and Amini, M. 2015. Antibiotic susceptibility of Clostridium perfringens from organic broiler chickens. *Online Journal of Veterinary Research*, 19(7), 465–70.

Allen, J. and Bellini, J. 2017. Reducing antimicrobial use: A practitioner experience. *In Practice*, 39, 462–73.

Alvarez-Fernandez, E., Dominguez-Rodriguez, J., Capita, R. and Alonso-Calleja, C. 2012. Influence of housing systems on microbial load and antimicrobial resistance patterns of *Escherichia coli* isolates from eggs produced for human consumption. *Journal of Food Protection*, 75, 847–53.

Alvarez-Fernandez, E., Cancelo, A., Diaz-Vega, C., Capita, R. and Alonso-Calleja, C. 2013. Antimicrobial resistance in *E. coli* isolates from conventionally and organically reared poultry: A comparison of agar disc diffusion and Sensi Test Gram-negative methods. *Food Control*, 30, 227-34.

Bahrndorff, S., Rangstrup-Christensen, L. Nordentoft, S. and Hald, B. 2013. Foodborne disease prevention and broiler chickens with reduced campylobacter infection. *Emerging Infectious Diseases*, 19(3), 425-30.

Berge, A. C., Hancock, D. D., Sischo, W. M. and Besser, T. E. 2010. Geographic, farm, and animal factors associated with multiple antimicrobial resistance in fecal *Escherichia coli* isolates from cattle in the western United States. *Javma-Journal of the American Veterinary Medical Association*, 236, 1338-44.

Brennan, M. and Christley, R. 2012. Biosecurity on cattle farms: A study in North-West England. *PLoS ONE*, 7(1), e28139.

Busin, V., Kenyon, F., Parkin, T., McBean, D., Laing, N., Sargison, N. and Ellis, K. 2014. Production impact of a targeted selective treatment system based on liveweight gain in a commercial flock. *Veterinary Journal*, 200(2), 248-52.

Cicconi-Hogan, K. M., Belomestnykh, N., Gamroth, M., Ruegg, R. L., Tikofsky, L. and Schukken, Y. H. 2014. *Short communication*: Prevalence of methicillin resistance in coagulase-negative staphylococci and *Staphylococcus aureus* isolated from bulk milk on organic and conventional dairy farms in the United States. *Journal of Dairy Science*, 97, 2959-64.

De Vries, M., Kwakkel, R. P. and Kijlstra, A. 2006. Dioxins in organic eggs: A review. *Njas-Wageningen Journal of Life Sciences*, 54, 207-21.

EFSA. 2016. The European Union summary report on trends and sources of zoonoses, zoonotic agents and food-borne outbreaks in 2015. European Food Safety Authority and European Centre for Disease Prevention and Control. doi:10.2903/j. efsa.2016.4634. Accessed at: https://www.efsa.europa.eu/en/efsajournal/pub/4634.

Ellis, K. A., Billlington, K., McNeil B. and McKeegan D. E. F. 2009. Public opinion on UK milk marketing and dairy cow welfare. *Animal Welfare*, 18, 267-82.

Ellis, K.A., Jackson, A., Bexiga, R., Matthews, J., McGoldrick, J., Gilleard, J. and Forbes, A. B. 2011. Use of diagnostic markers to monitor fasciolosis and gastrointestinal nematodes on an organic dairy farm. *Veterinary Record*, 169(20), 524. doi:10.1136/vr.d5021.

Fox, N., White, P. C. L., McClean, C. J., Marion, G., Evans, A. and Hutchings, M. R. 2011. Predicting impacts of climate change on *Fasciola hepatica* risk. *PLoS ONE*, 6(1), e16126.

Fraqueza, M. J., Martins, A., Borges, A. C., Fernandes, M. H., Fernandes, M. J., Vaz, Y., Bessa, R. J. B. and Barreto, A. S. 2014. Antimicrobial resistance among *Campylobacter* spp. strains isolated from different poultry production systems at slaughterhouse level. *Poultry Science*, 93, 1578-86.

Galis, A. M., Van, I. and Théwis, A. 2012. Organoleptic, chemical and microbiological quality of table eggs obtained in different exploitation systems for laying hens in Romania. Scientific Papers, Series D. *Animal Science*, 55, 162-6.

Gerzova, L., Babak, V., Sedlar, K., Faldynova, M., Videnska, P., Cejkova, D., Jensen, A. N., Denis, M., Kerouanton, A., Ricci, A., Cibin, V., Osterberg, J. and Rychlik, I. 2015. Characterization of antibiotic resistance gene abundance and microbiota composition in feces of organic and conventional pigs from four EU countries. *PLoS ONE*, 10, e0132892.

Guarddon, M., Miranda, J. M., Rodriguez, J. A., Vazquez, B. I., Cepeda, A. and Franco, C. M. 2014. Quantitative detection of tetracycline- resistant microorganisms in conventional and organic beef, pork and chicken meat. *Cyta-Journal of Food*, 12, 383–8.

Hanning, I., Biswas, D., Herrera, P., Roesler, M. and Ricke, S. C. 2010. Prevalence and characterization of *Campylobacter jejuni* isolated from pasture flock poultry. *Journal of Food Science*, 75, M496–502.

Hardy, B., Crilly, N., Pendleton, S., Andino, A., Wallis, A., Zhang, N. and Hanning, I. 2013. Impact of rearing conditions on the microbiological quality of raw retail poultry meat. *Journal of Food Science*, 78, M1232–5.

Helke, K. L., McCrackin, M. A., Galloway, A. M., Poole, A. Z., Salgado, C. D. and Marriott, B. P. 2016. Effects of antimicrobial use in agricultural animals on drug-resistant foodborne salmonellosis in humans: A systematic literature review. *Critical Reviews in Food Science and Nutrition*, 57, 472–88.

Hoorebeke, S. V., Dewulf, J., Immerseel, F. V. and Jorgensen, F. 2012. Production systems for laying hens and broilers and risk of human pathogens. In: Sandilands, V. and Hocking, P. (Eds), *Proceedings of the 30th Poultry Science Symposium*. University of Strathclyde, Glasgow, Scotland, 2011, pp. 77–96. doi:10.1079/9781845938246.0077.

Hovi, M. 2005. Animal health security – is it important on organic farms? *Proceedings of the SAFO Workshop*, September 2004, Falenty, Poland, pp. 4–14.

Huijbers, P. M. C., Van Hoek, A. H. A. M., Graat, E. A. M., Haenen, A. P. J., Florijn, A., Hengeveld, P. D. and Van Duijkeren, E. 2015. Methicillin-resistant *Staphylococcus aureus* and extended-spectrum and AmpC beta-lactamase-producing *Escherichia coli* in broilers and in people living and/or working on organic broiler farms. *Veterinary Microbiology*, 176, 120–5.

Jackson, A. 2012. Parasitic Gastroenteritis in calves during their first season at grass on organic and conventional farms in Scotland: The potential for a performance-based targeted selective treatment anthelmintic treatment programme. PhD Thesis, University of Glasgow, Glasgow, Scotland.

Kassem, I. I., Kehinde, O., Kumar, A. and Rajashekara, G. 2017. Antimicrobial-resistant *Campylobacter* in organically and conventionally raised layer chickens. *Foodborne Pathogens and Disease*, 14, 29–34.

Laanen, M., Persoons, D., Ribbens, S., De Jong, E., Callens, B., Strubbe, M., Maes, D. and Dewulf, J. 2013. Relationship between biosecurity and production/antimicrobial treatment characteristics in pig herds. *The Veterinary Journal*, 198(2), 508–12.

Leifert, C., Ball, K., Volakakis, N. and Cooper, J. M. 2008. Control of enteric pathogens in ready-to-eat vegetable crops in organic and 'low input' production systems: A HACCP-based approach. *Journal of Applied Microbiology*, 41, 186–9.

Luzardo, O. P., Rodriguez-Hernandez, A., Quesada-Tacoronte, Y., Ruiz-Suarez, N., Almeida-Gonzalez, M., Alberto Henriquez-Hernandez, L., Zumbado, M. and Boada, L. D. 2013. Influence of the method of production of eggs on the daily intake of polycyclic aromatic hydrocarbons and organochlorine contaminants: An independent study in the Canary Islands (Spain). *Food and Chemical Toxicology*, 60, 455–62.

MacLachlan, N. J. and Mayo, C. E. 2013. Potential strategies for control of bluetongue, a globally emerging, Culicoides-transmitted viral disease of ruminant livestock and wildlife. *Antiviral Research*, 99, 79–90.

Malissiova, E., Papadopoulos, T., Kyriazi, A., Mparda, M., Sakorafa, C., Katsioulis, A., Katsiaflaka, A., Kyritsi, M., Zdragas, A. and Hadjichristodoulou, C. 2017. Differences in sheep and goats milk microbiological profile between conventional and organic farming systems in Greece. *The Journal of Dairy Research*, 84(2), 206–13.

Meerburg, B. G. and Borgsteede, F. H. M. 2010. Organic agriculture and its contribution to zoonotic pathogens. In: Krause, D. and Hendrick, S. (Eds), *Zoonotic Pathogens in the Food Chain*. CABI International, Wallingford, UK, pp. 167–81.

Mellor, D. J. 2016. Updating animal welfare thinking: Moving beyond the 'Five Freedoms' towards 'A Life Worth Living'. *Animals*, 6, 21. doi:10.3390/ani6030021.

Osterberg, J., Wingstrand, A., Jensen, A. N., Kerouanton, A., Cibin, V., Barco, L., Denis, M., Aabo, S. and Bengtsson, B. 2016. Antibiotic resistance in *Escherichia coli* from pigs in organic and conventional farming in four European countries. *PLoS ONE*, 11, e0157049.

Padel, S. and Foster, C. 2005. Exploring the gap between attitudes and behaviour: Understanding why consumers buy or do not buy organic food. *British Food Journal*, 107, 606–25.

Park, Y. K., Fox, L. K., Hancock, D. D., McMahan, W. and Park, Y. H. 2012. Prevalence and antibiotic resistance of mastitis pathogens isolated from dairy herds transitioning to organic management. *Journal of Veterinary Science*, 13, 103–5.

Peng, M., Salaheen, S., Almario, J. A., Tesfaye, B., Buchanan, R. and Biswas, D. 2016. Prevalence and antibiotic resistance pattern of *Salmonella* serovars in integrated crop-livestock farms and their products sold in local markets. *Environmental Microbiology*, 18, 1654–65.

Petersen, H. H., Wang, J., Katakam, K. K., Mejer, H., Thamsborg, S. M., Dalsgaard, A., Olsen, A. and Enemark, H. L. 2015. *Cryptosporidium* and *Giardia* in Danish organic pig farms: Seasonal and age-related variation in prevalence, infection intensity and species/genotypes. *Veterinary Parasitology*, 214, 29–39.

Piskorska-Pliszczyska, J., Strucinski, P., Mikolajczyk, S., Maszewski, S., Rachubik, J. and Warenik-Bany, M. 2015. Dioxins, furans and PCBs in eggs: Results of 2006-2014 surveys. *Medycyna Weterynaryjna-Veterinary Medicine-Science and Practice*, 71, 696–705.

Salaheen, S., Chowdhury, N., Hanning, I. and Biswas, D. 2015. Organic poultry production with natural feed supplements as antimicrobials zoonotic bacterial pathogens and mixed crop-livestock farming. *Poultry Science*, 94, 1398–410.

Salaheen, S., Peng, M. and Biswas, D. 2016. Ecological dynamics of *Campylobacter* in integrated mixed crop-livestock farms and its prevalence and survival ability in post-harvest products. *Zoonoses and Public Health*, 63, 641–50.

Sancheza, H. M., Echeverria, C., Thulsiraj, V., Zimmer-Faust, A., Flores, A., Laitz, M., Healy, G., Mahendra, S., Paulson, S. E., Zhu, Y. and Jay, J. A. 2016. Antibiotic resistance in airborne bacteria near conventional and organic beef cattle farms in California, USA. *Water Air and Soil Pollution*, 227, 280.

Santman-Berends, I. M. G. A., Gonggrijp, M. A., Hage, J. J., Heuvelink, A. E., Velthuis, A., Lam, T. J. G. M. and Van Schaik, G. 2017. Prevalence and risk factors for extended-spectrum beta-lactamase or AmpC-producing *Escherichia coli* in organic dairy herds in the Netherlands. *Journal of Dairy Science*, 100, 562–71.

Sapkota, A. R., Hulet, R. M., Zhang, G. Y., McDermott, P., Kinney, E. L., Schwab, K. J. and Joseph, S. W. 2011. Lower prevalence of antibiotic-resistant enterococci on us conventional poultry farms that transitioned to organic practices. *Environmental Health Perspectives*, 119(11), 1622–8.

Sapkota, A. R., Kinney, E. L., George, A., Hulet, R. M., Cruz-Cano, R., Schwab, K. J., Zhang, G. and Joseph, S. W. 2014. Lower prevalence of antibiotic-resistant Salmonella on large-scale US conventional poultry farms that transitioned to organic practices. *Science of the Total Environment*, 476, 387–92.

Schwaiger, K., Schmied, E. M. V. and Bauer, J. 2010. Comparative analysis on antibiotic resistance characteristics of *Listeria* spp. and *Enterococcus* spp. isolated from laying hens and eggs in conventional and organic keeping systems in Bavaria, Germany. *Zoonoses and Public Health*, 57, 171–80.

Silverlas, C. and Blanco-Penedo, I. 2013. *Cryptosporidium* spp. in calves and cows from organic and conventional dairy herds. *Epidemiology and Infection*, 141, 529–39.

Slany, M., Reslova, N., Babak, V. and Lorencova, A. 2016. Molecular characterization of *Toxoplasma gondii* in pork meat from different production systems in the Czech Republic. *International Journal of Food Microbiology*, 238, 252–5.

Suriyasathaporn, W. 2010. Milk quality and antimicrobial resistance against mastitis pathogens after changing from a conventional to an experimentally organic dairy farm. *Asian-Australasian Journal of Animal Sciences*, 23, 659–64.

Udikovic-Kolic, N., Wichmann, F., Broderick, N. A. and Handelsman, J. 2014. Bloom of resident antibiotic-resistant bacteria in soil following manure fertilization. *Proceedings of the National Academy of Sciences of the United States of America*, 111, 15202–7.

Wallander, C., Frossling, J., Dorea, F. C., Uggla, A., Vagsholm, I. and Lunden, A. 2016. Pasture is a risk factor for *Toxoplasma gondii* infection in fattening pigs. *Veterinary Parasitology*, 224, 27–32.

Webster, A. J. F. 2001. Farm animal welfare: The five freedoms and the free market. *The Veterinary Journal*, 161, 229–37.

Williams, P. R. D. and Hammit, J. K. 2001. Perceived risks of conventional and organic produce: Pesticides, pathogens, and natural toxins. *Risk Analysis*, 21(2), 319–30.